John Gray McKendrick

Life in motion: Muscle and nerve

A course of six lectures delivered before a juvenile auditory at the Royal institution

of Great Britain during the Christmas holidays of 1891-93

John Gray McKendrick

Life in motion: Muscle and nerve
A course of six lectures delivered before a juvenile auditory at the Royal institution of Great Britain during the Christmas holidays of 1891-93

ISBN/EAN: 9783337291150

Printed in Europe, USA, Canada, Australia, Japan

Cover: Foto ©berggeist007 / pixelio.de

More available books at **www.hansebooks.com**

LIFE IN MOTION

OR

MUSCLE AND NERVE

A COURSE OF SIX LECTURES

Delivered before a Juvenile Auditory at the Royal Institution of Great Britain during the Christmas Holidays of 1891-92

BY

JOHN GRAY McKENDRICK

M.D., LL.D., F.R.S., F.R.C.P.E.

PROFESSOR OF PHYSIOLOGY IN THE UNIVERSITY OF GLASGOW AND FORMERLY FULLERIAN PROFESSOR OF PHYSIOLOGY IN THE ROYAL INSTITUTION OF GREAT BRITAIN

LONDON AND EDINBURGH

ADAM AND CHARLES BLACK

1892

W. J. M.

A dear son, full of youthful enthusiasm for science, suggested the title of this course of lectures during a walk we had among the hills in the autumn of last year, when the bloom was on the heather. He has since passed away, but it soothes me to dedicate to his memory this little book.

PREFACE

Although these lectures were delivered without the aid of a manuscript, they have been printed substantially as they were spoken, making allowance for the exigencies of an experimental lecture. Here and there I have introduced a little new matter to give some measure of completeness to the exposition. My object was to interest and instruct, and an appeal was made to experiment wherever that was possible. It is no doubt difficult to present the facts of science in a book with the same vividness as when they are demonstrated by experiment; but I have endeavoured, by introducing numerous illustrations, to suggest to the mind what was actually done at the lecture. The delivery of these lectures was a pleasure to myself, as, with the resources of the Royal Institution, I saw certain physiological phenomena more clearly than is possible even in a well-equipped laboratory, and I hope that their perusal will interest the young,—and even those who may feel they are no longer young,—in the aims and methods of physiological science.

UNIVERSITY OF GLASGOW,
April 1892.

CONTENTS

LECTURE I

LECTURE II

LECTURE III

LECTURE IV

LECTURE V

LECTURE VI

LIST OF ILLUSTRATIONS

LECTURE I

Introduction—Movement in general—Molecular movements—Muscles—Organ and function—Muscular contraction.

THE object of the courses of Christmas lectures at the Royal Institution is to interest the young in the principles and the progress of science. This has been many times successfully accomplished by describing in simple language and, if possible, by demonstrating, the laws that govern a well-known phenomenon, such as the burning of a candle or the formation of a soap-bubble. The study of these familiar things is the door by which we may enter into the domain of natural philosophy. As described by Faraday, a candle-flame became a centre around which we found clustered the fundamental facts and principles of chemical, and even of physical science; and the consideration of a soap-bubble (as in the lectures in this place by Professor Dewar two years ago), its production, form, colour, is worthy of the intellect of the most profound philosopher, while it is an unfailing source of amusement and instruction to the youthful mind.

But, my young friends,—and it is to you I shall address myself in these lectures, although I am glad to see so many present who are older in years, but who can still relish a simple lecture to juveniles,—there is another department of nature that is not, strictly speaking, the province of the natural philosopher. There are the phenomena of living matter, the events that happen in the life-history of living beings, the changes that occur in our own bodies, and on which our lives depend. These are investigated by the physiologist; and we shall find, as we go on, that they are not so easily demonstrated as many of the phenomena that happen in dead matter, and that they are, on the whole, more difficult to understand. Following, however, the example of previous lecturers, I have endeavoured to choose a subject, the consideration of which will form an introduction to physiology, which will illustrate how physiologists work in their laboratories, and how they reason about the problems they have to solve.

I have called the subject of the course "Life in Motion." We must take care at the outset not to get into difficulties about what is implied by the word Life. We shall not inquire, in the meantime at all events, whether life may be considered as something independent of matter, or whether it is the outcome of material arrangements. It will be better at first to use the words Living Matter instead of the word Life, and to define our subject of study as Living Matter in Motion, or the Motions of Living Matter. We might study other aspects of living matter, as, for example, the chemical changes occurring in it, the forms it assumes, its arrangements for different purposes, or its wonderful connection in some conditions, as in the brain, with consciousness. We shall, however, limit ourselves to the consideration of its motions; and we shall refer to the other properties of living matter only in so far as these throw light on the secret mechanism by which it moves.

We are familiar with motion every day of our lives. We know, first of all, of great movements of matter, such as the wheeling of the planets in their elliptical orbits round the sun, the spinning of the earth on its axis, or the still grander movements of the firmament, as revealed by the proper motions of the stars. Such movements impress the imagination with a sense of vastness and of irresistible power. Then there are the movements on the surface of the world itself—the tides, the flow of rivers, the hurricane, the clouds travelling athwart the sky, and many others. These are movements of great masses of matter; and when they are studied by the natural philosopher, he finds that they are regulated by well-known dynamical laws. All such movements are evident to the senses; but there are other movements that are not so, and which can be detected only by special methods of research. Such are those that occur, as it were, below the surfaces of things. These are called *molecular*, because the bodies that move are minute particles or molecules of matter, far too small to be seen with the eye or even with the aid of the most powerful microscope. Still, the natural philosopher tells us that the movements of these little particles are controlled by dynamic laws as definite in their operation as those that govern the planets in their journeys round the sun. The physicist, however, deals with another and more subtile class of movements in an Ether, which he supposes to pervade space. Waves, strains, pressures, whirls in the ether, account to men of science for many of the facts of heat, light, and electricity. Thus, according to the scientific conceptions of the present day, we have to imagine all the matter in the universe as in a state of movement, some movements large and occurring in vast stretches of space, and others almost inconceivably minute. Nothing in this vast mechanism has come to rest. Each particle of matter is quivering, molecules of all gases are vibrating to and fro, and millions of wavelets are streaming through the ether in all conceivable directions. If we suppose that the essence of life is movement, does not this give one a conception that in a sense the universe is alive?

We have to deal, however, in these lectures with the movements occurring in living matter. We all know that the living things with which we are familiar move. They move their bodies as a whole, or they move parts of their bodies. Animals run, leap, swim, fly, and perform many other movements. These movements are obvious. Every one can see them. But here, again, we must go a little farther in thought and look below the surface. We then find that in living things there are also molecular movements, and that the larger movements which we can all see depend on the small ones that are invisible to our eyes.

To aid us in understanding these statements a little better let us now perform a few experiments. Take, first of all, one or two chemical reactions. I add a solution of nitrate of silver to a solution of chloride of sodium. You see a solid matter appear, which soon falls to the bottom of the glass. This solid matter is chloride of silver, a new substance formed by the interaction of the nitrate of silver with the chloride of sodium. The change may be expressed in this way :—

Nitrate of Silver	+	Chloride of Sodium
consists of Nitric Acid and Silver		consists of Chlorine and Sodium
	gives	
Chloride of Silver	and	Nitrate of Soda
Chlorine and Silver.		Nitric acid and Sodium.

In other words, both substances—nitrate of silver and chloride of sodium—were split up into their constituents, and the molecules of chlorine, forsaking the sodium, travelled to and united with the molecules of silver to form a new substance called chloride of silver. Again, if I add a solution of iodide of potassium to a solution of corrosive sublimate, we get a beautiful coloured substance called iodide of mercury. A new substance has been formed by the iodine and potassium parting company, the iodine then uniting with the mercury of the corrosive sublimate to form the salt called iodide of mercury, while the potassium united with the chlorine to form a substance known as chloride of potassium. We may picture the molecules of these substances, at the moment of chemical change, rushing from one to another so as to form new combinations.

Look at what is going on in this large glass vessel. A few hours ago I placed in the vessel a solution of grape sugar, and I added a small quantity of fresh yeast. The fluid was at first clear, now you observe it is turbid and is yellow in colour. Notice also the froth gathering on the surface. Fermentation is going on actively in the fluid. Gas is passing off in large quantities. Under the action of the minute living vegetable cells that constitute yeast, the sugar is being split up into alcohol and carbonic acid. The alcohol remains in the fluid and the carbonic acid escapes into the air. Observe when I put this lighted taper into the vessel that the flame is at once extinguished

FIG. 1.—Yeast cells. The cells of *Saccharomyctes cerevisiæ*, multiplying by budding. Magnified 300 diameters.

by the carbonic acid gas. Here we have an example of molecular movements brought about by the action of living organisms—the yeast cells. Each little yeast cell acts directly or indirectly on the sugar, effects a decomposition, as I have said, into alcohol and carbonic acid, while other substances, such as glycerine and succinic acid, are formed in smaller quantity. During the fermentative process the temperature of the fluid rises, and the yeast cells grow and multiply, living, as it were, on the sugar and other nutritive matters in the fluid.

Let us pass to another experiment. I have prepared a saturated solution of acetate of sodium—that is, a solution which cannot be made stronger. You observe it is a clear fluid like water. I now drop in a crystal of acetate of sodium. You see at once crystals shooting through the fluid, and in a few moments the mass in the flask has become solid. The flask has also become perceptibly hotter. The agitation excited by dropping in the crystal has caused a rapid change in the position of the particles, the solution passing from the fluid to the solid state with the evolution of heat. This is another example of a molecular movement.

Consider next an experiment in which a state of movement can be appreciated by the eye. Look at the limbs of this large tuning-fork. You observe they are stationary. I strike the fork, and you see it is at once thrown into a state of vibration, as shown by the fuzzy appearance of the limbs when placed in the electric beam. Listen, and you hear a low, humming sound. This is caused by the movements of the fork sending a number of pulsations through the air, which strike against and agitate the drum-head of the ear, and from it the movements are communicated to the

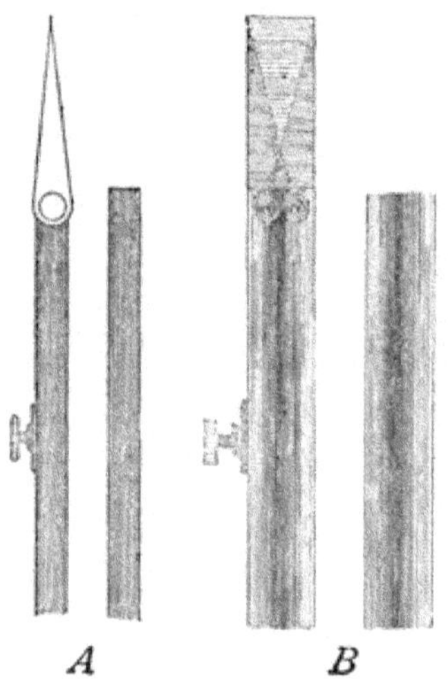

FIG. 2.—Movements of limbs of tuning-fork. A, limbs at rest; B, in movement.

nervous structures in the deeper ear. If I touch the limb of the fork with my finger, the fuzziness vanishes, the image now appearing sharply on the screen, and the sound is no longer heard. I shall next cause this smaller fork to sound by drawing a fiddle bow across one of its prongs. You hear it sounding. I place it in the electric beam. You see the shadow of its limbs are well defined on the screen. You cannot see its movements, as you saw the movements of the first fork, because they are too fast for the eye. But they are not too fast for the ear, because we hear the sound; and they are not too fast for the sense of touch, because when I touch the fork I feel a thrill against my fingers. The pressure of my finger stops the vibrations and we no longer hear the sound. The large fork vibrated 128 times in each second, and the smaller one moved twelve times as fast, or 1536 times a second.

I strike with a key a little cylinder of steel suspended on this support. You hear the sound, but I need hardly point out you cannot see the movements of the cylinder, nor do I feel a thrill when I touch it. The movements, occurring about 12,000 times a second, are too fast for the eye and too fast for touch, but they can still be followed by the ear. Lastly, I strike this smaller cylinder. You hear the dull thud of the stroke, but no piercing tone is heard; and yet this cylinder is no doubt vibrating, but its movements are too fast to be followed by any of the senses. Thus we learn that our senses are limited organs as regards the detection of movement. We can only follow periodic movements through a narrow

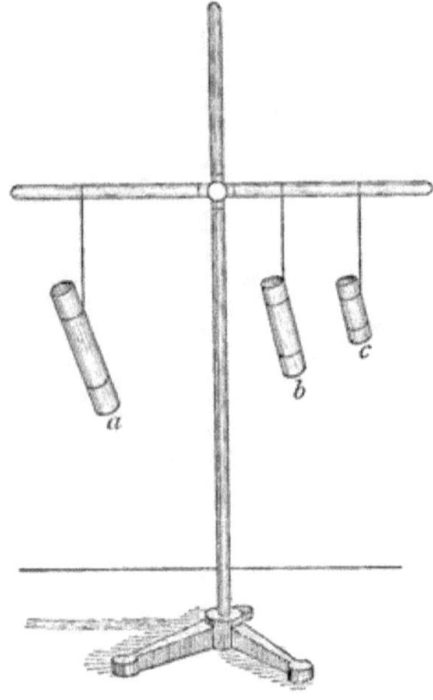

FIG. 3.—Steel cylinders emitting a high tone when struck. *a*, 12,000 vibrations, *b*, 20,000 vibrations, and *c*, 30,000 vibrations per second.

range; and there are regions in which delicate movements occur, which we can only explore by indirect methods and by processes of thought.

Let me point out to you, in passing, that man's supremacy over the lower animals lies in the power he possesses of pushing his inquiries far beyond the range of his senses. From facts that appeal to his senses he reasons as to phenomena that can never be directly observed, and by intellectual processes he can acquire knowledge as accurate as if he were able to examine the phenomena with organs of sense having powers much more extensive than those he possesses. A recognition of this quality of man's mind indicates also the value of education in science. This does not consist, as is often erroneously supposed, in merely acquiring a knowledge of fact, but also and more in learning to reason correctly and in cultivating the use of the imagination. The scientific thinker has a mental vision into regions far beyond the limited powers of his senses, and hence there is much truth in the statement that the greatest scientific men have many of the qualities of mind of the poet or of the prophet. They are seers in a true sense of the word.

Let us take another simple experiment or two from the region of physics, to prepare the way for our own special study. Here is a mass of soft iron, like the half of a link of a large chain. Copper wire has been coiled round each limb, and we can connect the ends of the copper wire with an electric battery. If I take into my hands the ends of the wires coming from the battery, I feel nothing, so that any force that may be coming from the battery is not

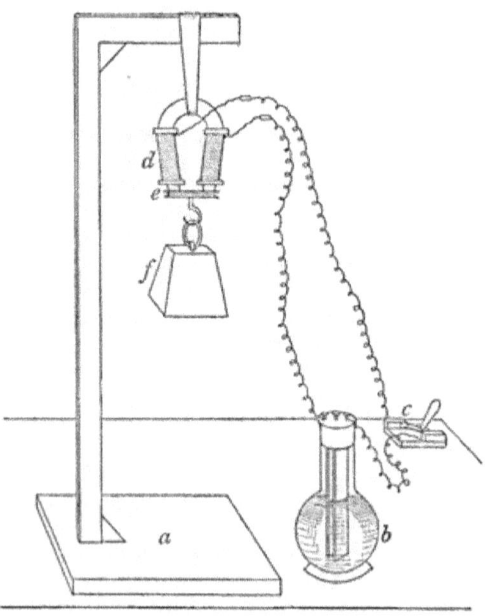

FIG. 4.—Electro-magnet supporting a weight. *a*, stand; *b*, galvanic element; *c*, key; *d*, electro-magnet; *e*, keeper; *f*, weight.

affecting my skin so as to cause any kind of sensation; but if I touch the tip of my tongue with the wires, I have a peculiar sensation of taste. The tongue tells me that something is affecting me, although the hands are apparently not influenced. Still we know that chemical changes are going on in this battery, and that these changes are followed by something we call a current of electricity. The best proof of this I can give you is that if I send this current round the wire covering the ends of the link of soft iron, you will find the properties of the soft iron altered so that it becomes powerfully magnetic. You observe that when no current is flowing, as is the case when I break the connection, the soft iron has no attractive influence on this piece of steel; but when I send the current on, at once the soft iron becomes so powerfully magnetic that it attracts the steel keeper with great force—a force so great that you see the keeper supporting a heavy weight. A wonderful change has been wrought in the soft iron—a change depending on molecular movements far too fine to come within the range of direct observation. It can be shown, however, that such a piece of soft iron actually elongates when the current passes round it, as in this experiment; and if I were to interrupt this current so as to send it at short intervals of time, the mass of soft iron would vibrate so as to give out a musical tone. It is not in my province to discuss these remarkable physical phenomena; but I ask you to remember this simple and familiar experiment, because it not only is an example of what we mean by molecular movement, but it explains the construction of many physiological appliances we shall use in future lectures.

FIG. 5.—Muscles in human arm. *b*, biceps muscle.

Now we come to movements associated with life. These are best studied in a muscle. Flex your arm at the elbow joint, and you will feel the flesh above the joint and in front of the arm become firm and hard. Extend the arm and it again becomes soft. These movements are made by the action of special organs we call muscles; and the particular muscle I ask you to notice in the arm is called the biceps. Every boy knows where his biceps muscle is, and when he is training for throwing the cricket ball, or for rowing, he feels his stiff and firm biceps with a certain amount of commendable pride. A young lady has a biceps also;

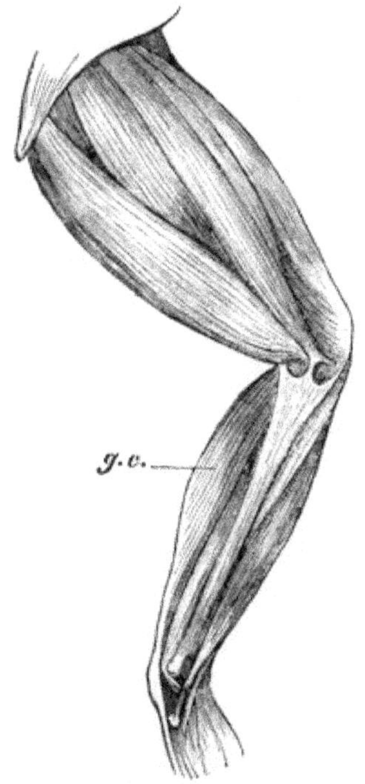

FIG. 6.—Muscles of frog's leg. *g, c*, the gastrocnemius muscle often used in physiological experiments.

but she is not so much interested in its welfare. In a section of a limb, such as you find at any butcher's stall, you see the masses of red flesh imbedded in fat and loose tissue. These masses of red flesh are the muscles. But if we dissect a limb, we find that the muscles are beautiful organs, adapted, as regards form and length and bulk, to the work each has to perform. This diagram will give you a conception of what we mean by a muscle. It shows the muscles of a frog's leg. You observe that, as a rule, a muscle springs from a bone, and is attached at the other end to another bone, a joint, sometimes two or more joints, intervening. One end of such a muscle as the gastrocnemius (Fig. 6) muscle terminates in what is called a tendon or sinew —a fibrous structure which is attached to a membrane covering the bone. The muscles are the living ropes that pull the parts of the animal machine.

We may call the muscles the organs of movement. By an organ physiologists mean a part of the body devoted to a special use or purpose, or, as we say, a function. A muscle has its own work to do, in a sense as true as that the heart has its own work to do in acting as a force-pump to drive on the blood through the blood-vessels, or, in other words, to keep up the circulation. It is important to notice that a muscle may be regarded as an independent living thing, an organ that in certain conditions might live and work by itself. It has its own blood-vessels for supplying it with nourishment, its own nerves for stimulating it to activity, or for putting it into relation with the headquarters in the nervous system, and its own arrangements for the removal of so-called waste matters that have arisen from the tear and wear of the muscle in its active life.

Now, suppose we could isolate a muscle from the rest of the body and keep it alive, you can see that we might be able to examine the changes that occur in it when it works. So long as it is in the body, we cannot easily subject it to the method of experimental inquiry, because it is, in the first place, part of a living sentient being; and, in the next place, it is part of a complicated organism, the functions of which are all so closely connected, that if we interfere with the mechanism of one part we interfere with the whole, and this disturbance of the functions of the body as a whole reacts upon the functions of the very part we desire to study. Obviously, then, our course is to remove the muscle from the body and to keep it alive. But we are met by the difficulty that if we do so the muscle will soon die. Here you see at once one of the serious disadvantages at which the physiologist is placed in the prosecution of his science. The things the natural philosopher deals with are dead things. He can isolate them and interrogate them at pleasure by experiment, submitting them to all sorts of conditions and changes of circumstances, without the risk of destroying them or even of altering the property he wishes to examine. Many of the phenomena he has to investigate are of a tolerably permanent character, and the things he operates upon are, as a rule, not the seat of constant change. I admit that this is only generally true, and that there are phenomena sometimes investigated by the physicist which are almost, if not quite, as brief and evanescent as those that come under the eye of the physiologist. Still the statement is true in the main.

The things the physiologist has to investigate only work within a narrow range of conditions. Alter the blood supply, allow drying to take place, change the temperature, and the phenomenon he is in search of cannot be found. Modify the conditions of life beyond certain limits, and death at once begins. In trying to find out what are the phenomena of life we arrest the very phenomena we are in search of. Thus we kill the goose that lays the golden eggs.

Look at this piece of clockwork. Suppose I saw it for the first time and desired to know how it worked. I could do so by watching the movements, observing the slow uncoiling of the chain, and the movements of pinion and toothed wheel. I might also take it to pieces and study the various parts. By taking it to pieces the mechanism no doubt would stop, but I might, by careful consideration of how one thing fitted into another, ascertain how the thing worked.

The body may be regarded as an extremely complex machine, intimately connected in all its parts, but yet it is possible to make out, by direct inspection, something about the uses of its individual parts. We can see that the skeleton forms a scaffolding for the soft parts of the body, that the muscles and joints form a system of levers by which movements are effected, that the heart pumps the blood, and that the lungs are used for breathing. We can stop the machine and study the form and structure of the various parts. This is the province of the anatomist. Further, as I have said, we can watch the machine in action. This is the work of the physiologist. But the peculiarity of the physiologist's machine is that each part of the mechanism is alive. Each individual organ is a machine or instrument by itself, and its use as a part of the whole complex machine depends on the molecular machinery which composes the individual organ. Returning to the analogy of the watch, it is as if each wheel, and pinion, and chain were a separate machine, more complex, perhaps, in structure than the watch itself. It is as if we had wheels within wheels, and as if the mechanism depended partly on the large wheels, and partly and mainly on the small wheels within the large ones.

In like manner our study of muscle must include the action of the muscle as a part of the body, and also the changes that happen in the muscle itself, and by which it works. We shall take the last part of the investigation first, and we shall therefore try to find out what a muscle does and how it works.

Let us interrogate a muscle itself. I have prepared the muscle of a frog in the way shown in this diagram. Now I wish you to follow all I do; and you must receive all the explanations that I would think it necessary to make

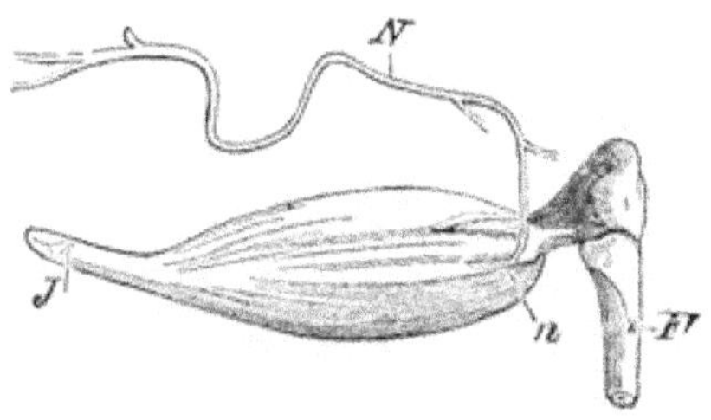

FIG. 7.—The gastrocnemius muscle of a frog prepared for experiment. F, femur, bone of the leg; N, sciatic nerve ending in muscle at *n*; J, tendon of Achilles, with a small hole in it for a hook.

if you were beside me in my laboratory. How has the muscle been prepared? As you know, the frog is what is usually called a cold-blooded animal, that is to say, the temperature of the body is always not far from that of the medium in which it lives. The term "cold-blooded" is misleading, because the frog's blood may, in some circumstances, not be cold; and besides it is a term to which one would think the frog might take just exception as being a term to which an evil meaning has been attached, as when we speak of a cold-blooded villain! The term "variable temperature" is more correct, as it distinguishes all such animals from those in which the blood has an almost uniform temperature, whatever may be the temperature of the medium in which the animal lives. For example, the temperature of a healthy man in the burning plains of India or in the snowy wastes of Siberia never varies much from 98·4° Fahrenheit.

It is one of the characteristics of animals of variable temperature, like the frog, that all its tissues are more stable or permanent than those of animals of uniform temperature. The tissues of a frog are not so liable to change as those of a rabbit or of a bird. Now the active phenomena of life all depend on instability of tissue. Imagine you have built a house of cards. You might build it so that the slightest push, or even a whiff of air, might cause it to fall to pieces; or you might so construct it that considerable force would be needed to knock it down. In the first case, the house of cards would be unstable; in the second, it would be stable. The tissues from the two kinds of animals differ in a similar way. When the death of the animal occurs, when breathing ceases and the blood stops flowing, in the case of an animal of uniform temperature the unstable tissues at once begin to change, and they speedily lose their vital properties and die. On the other hand, in similar circumstances, the more stable tissues of an animal of variable temperature undergo but little change for a considerable time. Thus it is that the tissues of a frog live much longer after the death of the animal than those of a rabbit, or of a rat, or of a man. The muscles live after the death of the animal. As an individual, the frog is dead. It died instantaneously and without pain, but its muscles still live, and, in suitable conditions, they may live for hours. Thus you see that the mysterious property we call life (if you choose to call it a property) is not in one part of the body more than in another, but is diffused through it. We have stopped the watch, but bits of the mechanism are still going on. By and by they will stop also, and then there will be complete death.

Now look at our preparation (see Fig. 7 p. 23). You see the muscle—the gastrocnemius we call it—attached to the lower end of the thigh-bone or femur. The other end terminates in a tendon—the tendon of Achilles —fixed to the heel of the foot. It is the same muscle as forms the calf of the leg in our own bodies, and the tendon is the firm band you can feel above the heel. Observe the whitish thread passing into the preparation. This is the sciatic nerve, a great nerve that runs down the back of the thigh, sending branches into the gastrocnemius and to other muscles. I shall now clamp the upper end of the femur or thigh-bone by these forceps and pass a little hook through the tendon. This hook is attached to a silk thread connected with this instrument, which we may call a muscle telegraph, and by which any movement of the muscle will be made visible to us. Then we stretch the nerve across two little platinum wires coming from an electric arrangement which I shall not at present explain (Fig. 8). We will not use the telegraph, as the signal may not be easily seen, but we will cause the muscle when it contracts

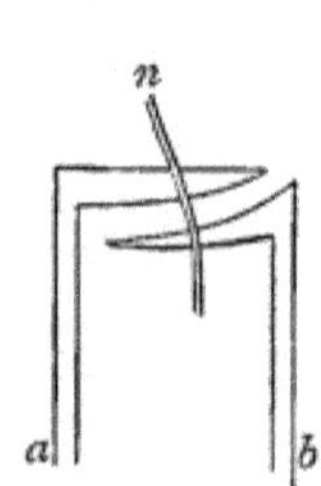

FIG. 8.—Platinum wires, *a* and *b*, with nerve, *n*, stretched across them.

to ring a little bell. Here you see the shadow of the apparatus cast on the screen by the electric light. I intend to use electricity to stimulate or irritate the nerve or the muscle at pleasure, and I have so arranged the apparatus that I

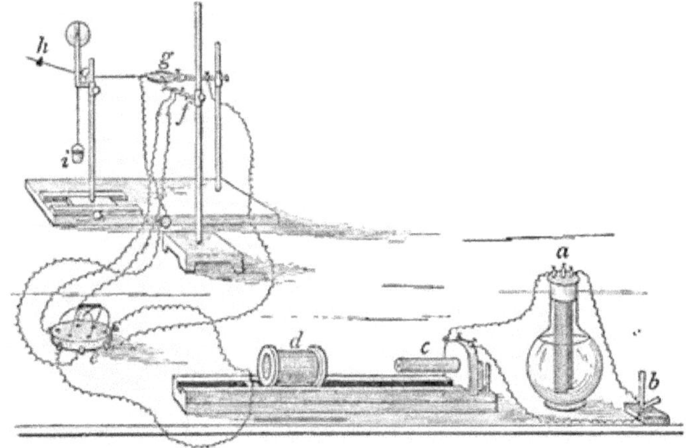

FIG. 9.—Arrangement of apparatus for irritating the muscle, *g*, directly, or the nerve connected with it at *f*. *a*, galvanic element; *b*, key; *c*, primary coil of induction machine; *d*, secondary coil; *e*, commutator by which current can be sent either to nerve or to muscle; *f*, electrodes, as in Fig. 8, for the nerve; *g*, muscle; *h*, small hammer; *i*, weight.

can send a single shock to the nerve or muscle, or a number of shocks in quick succession. We have now got everything adjusted. Observe that when I send a shock to the nerve the muscle gives a twitch and pulls on the thread, and observe the little hammer striking the bell. The muscle has contracted only for an instant and you hear one stroke of the bell. Another shock causes another twitch; and we find that if we allow some time to elapse between successive shocks, there is a twitch with each shock. But the twitch is so

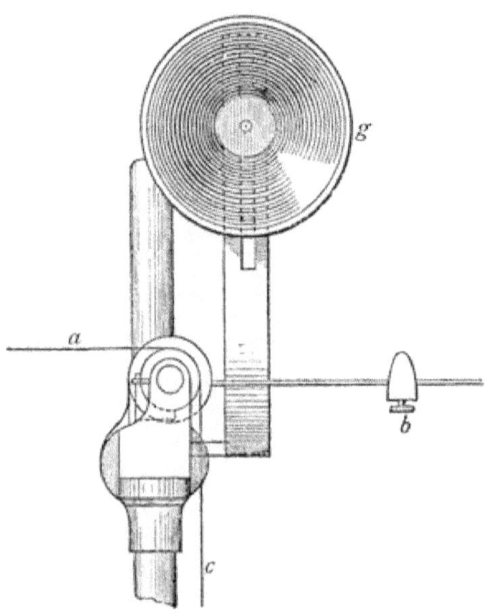

FIG. 10.—An enlarged view of the muscle-bell. *a*, thread coming from tendon of muscle; *c*, thread for weight; *b*, hammer for the bell *g*.

fast, it occurs in so short a time, that the eye can scarcely follow it, so that we cannot see what the muscle really does. But I now send a rapid series of shocks, and you observe that the muscle has become shorter and thicker. It has also pulled on the thread of the telegraph, and the hammer of the bell is kept up. This condition you will notice is not a rapid, sudden twitch, but a slow, steady, persistent pull or contraction. The muscle has passed into a state of cramp, or, as physiologists term it, a state of tetanus. The short, sudden contraction we shall call a twitch or simple spasm.

We repeat the experiment, only irritating the muscle instead of the nerve, and we get the same result : a single twitch with a single shock and tetanus when the shocks come in rapid succession.

We have learned from this experiment, then (1), that when we irritate the nerve going to a muscle, the muscle becomes shorter and thicker, or, in other words, it contracts ; (2), that a single shock of electricity to the nerve is followed by a sudden sharp twitch, a single contraction ; and (3) that a number of shocks sent in rapid succession to the nerve causes the muscle to pass into a more lasting state of contraction called tetanus or cramp.

In the experiment we have just performed all the work the muscle did was to pull up the signal and ring the bell. It does not require much energy to do this, and the experiment gives one a very inadequate notion of the amount of energy that can be brought into play by a small muscle like the one we are now studying. Here is another muscle of the same size. It weighs about half a gramme, or about seven grains. We have suspended it so that

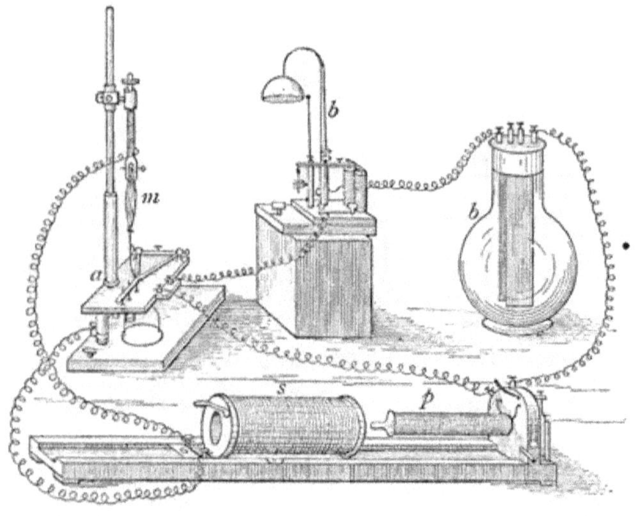

FIG. 11.—Arrangement of apparatus for showing muscle lifting a weight. *b*, galvanic element; *b* (in middle), electric bell; *p* primary and *s* secondary coil of induction machine; *a*, frog interrupter; *m*, muscle. See next figure.

when it contracts it lifts a lever, and, breaking an electric circuit, causes a bell to ring. This apparatus, I may mention, was sent to me for these lectures by Professor du Bois Reymond of Berlin, who lectured on "Nerve and Muscle" in this Institution in 1855, and who has taken a warm interest in the success of the present course. I put a weight on the scale-pan below the lever and irritate the muscle. Observe it can lift, as you hear by the tone of the bell, 5, 10, 20, 30, 40, 50, 100, 200, 250, 300, 350, 400, even up to 500 grammes. It can

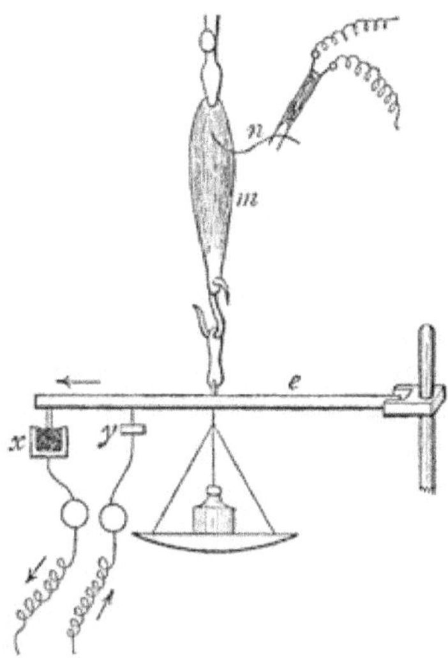

FIG. 12.—Essential part of frog interrupter used in experiment represented in Fig. 11. *m*, muscle; *n*, nerve; *e*, lever; when *e* is raised by the contracting muscle *m*, the contacts at *x* and *y* are broken; *x* is a platinum wire dipping into mercury, and *y* is a contact between two platinised surfaces. The arrows near the wires connecting *x y* show direction of current. When contact is broken by lifting the lever *e*, the bell *b* (middle) in Fig. 11 rings.

actually, by a sheer pull, move a mass one thousand times its own weight. Is not this a wonderful expenditure of mechanical energy?

The obvious phenomenon of a muscle, then, is that it contracts or changes its shape when it is irritated. We say that a muscle, like other living matter, is irritable. By this we mean that it responds to a stimulus, and the response is a change of form, a contraction. We shall see, however, that the contraction or change of shape is associated with many other changes of a molecular character not obvious to the senses but to be looked for by special methods. We must also, in next lecture, study more carefully the contraction itself.

LECTURE II

Myograph—Electricity as stimulus—Action of nerve—Continuous current—Interrupted current—Induction coil—Graphic method—Time in physiological phenomena—Chronographs—Study of contraction—Latent period.

In last lecture we saw that when the nerve connected with a muscle is irritated, the muscle changes its form, that is to say, it becomes shorter and thicker. It is this shortening or, as it is termed, contraction, that causes the movement of one part of the skeleton upon another. Let us repeat the experiment and study it more closely. This time we shall make use of an apparatus called a myograph or muscle-writer, and by means of it the muscle will write down its movement on a smoked-glass plate. The instrument is shown in this diagram. The upper end of the femur is fixed by the clamp C, sliding on the pillar B, and the tendo Achilles is attached by a hook to the horizontal bar E E, which carries a marker J; this marker is brought into contact

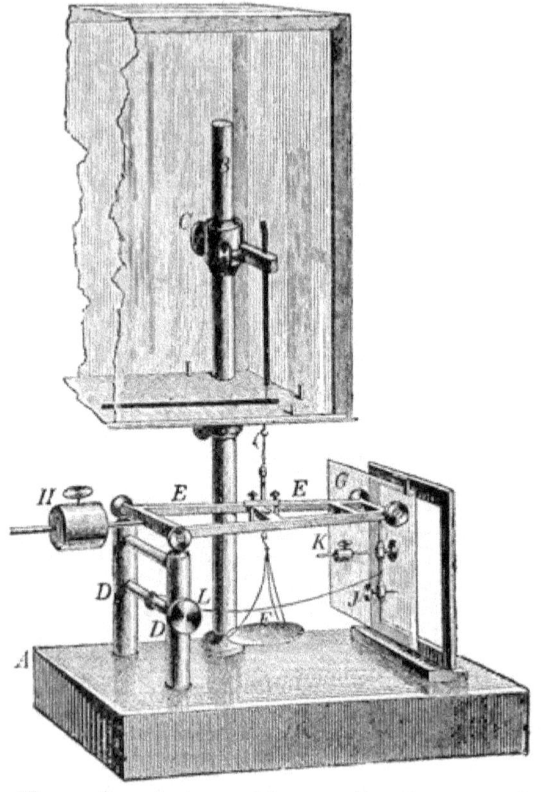

FIG. 13.—Myograph, an instrument for recording the contraction of a muscle. A, wooden stand; B, vertical brass pillar; C, sliding forceps or clamp for holding upper end of femur of nerve-muscle preparation; D D, short vertical pillars, on the top of which the lever E E works; F, scale-pan for weight; J, marker; G, glass plate; K, counterpoise to keep J in contact with G; H, counterpoise to lever E E. The nerve-muscle preparation is covered with a glass shade, to the walls of which pieces of wet blotting-paper are attached, thus forming a moist chamber to keep the nerve from drying.

with G, a smoked-glass plate that can be

horizontally moved, sliding in grooves. The nerve is stretched over wires coming from a battery or induction coil, so that it may be irritated by an electric current. When the nerve is irritated, you observe the muscle contracts, lifts up the lever E E, and the

FIG. 14.—Examples of tracings taken with the myograph shown in Fig. 13. Plate moved at intervals in direction of arrow.

marker J draws a vertical line on the plate G. We then push the plate a little farther on, and again stimulate the nerve by a shock. Another vertical line is drawn on the smoked plate; and, by repeating the experiment, a number of vertical lines can thus be drawn. Suppose we put a weight in the scale-pan F below the frame, the height of the line drawn on the smoked plate, making allowance for the increased amplitude of the movement obtained by the lever, will indicate the work done by the muscle in lifting the weight. We will see by and by that a muscle not only may do work by lifting a weight, but that it becomes hotter in doing so. Energy is thus set free from the muscle as mechanical energy and heat.

The first time one sees this experiment it is not easy to be satisfied that the electricity is used merely as a means of irritating the nerve going to the muscle, and that it is not the agent that causes the muscle to contract. Suppose a bit of nerve like a thread is stretched over two wires connected with our electric apparatus, as in this diagram on the blackboard, the current of electricity enters the nerve by the one wire and issues from it by the other. It only passes through a small bit of nerve, from *a* to *b*; it does not run down the nerve to the muscle, but, in passing along the bit of nerve, the nerve is irritated, a molecular change of some kind is generated in it, and this change travels down the nerve to the muscle in the direction of the arrow. The change that runs along the nerve, as the

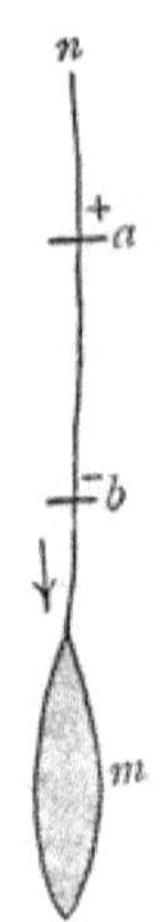

FIG. 15.—Diagrammatic representation of nerve, *n*, and muscle, *m*. Current of electricity enters at *a* and passes to *b*.

result of irritating it, we call the nerve-current, and it is this that excites the muscle into action. When the nerve-current reaches the muscle, it sets up molecular changes in it, and these changes are expressed to our eyes by a contraction, heat at the same time being liberated. The electric current we call a stimulus. We use electricity for stimulating the nerve because it is a convenient method, but the nerve, as we shall see, might be stimulated in other ways, as by mechanical irritation, such as pricking, pinching, or beating, or by heating it suddenly. It does not matter, however, how we stimulate the nerve; the result, so far as the muscle is concerned, is always the same: it always contracts.

But you will naturally ask, is there any relation between the strength of the current we employ for stimulating the nerve and the amount of work the muscle can do in lifting a weight? There is no direct quantitative relation. A very feeble current is quite sufficient to set the nerve in action, just as the pull of a hair trigger is enough to set free the energy from a charge of gunpowder in a rifle.

The nerve-current sets free the energy already stored in the muscle. The muscle substance is a magazine or store of energy, and this energy is set free by the molecular action of the nerve.

Having got hold of these general notions as

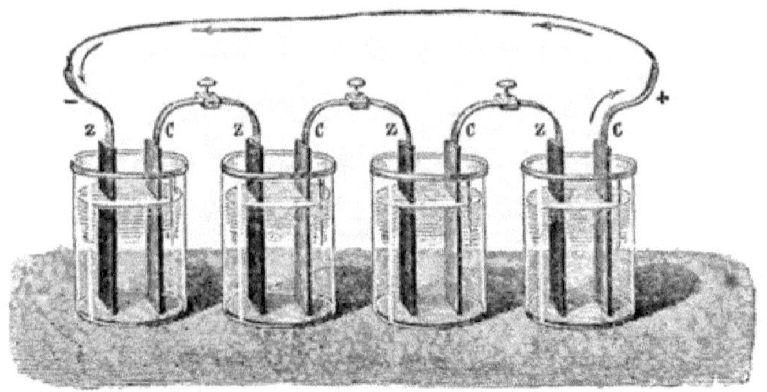

FIG. 16.—Battery of galvanic elements. Z, zinc plate; C, carbon plate. Arrows show direction of current.

to how a nerve acts on a muscle, let us consider for a little the nature of the electric shocks we employ for stimulating the nerve.

I have here a number of galvanic elements joined together so as to constitute what we term a battery. Electricity is generated in these elements in consequence of chemical changes occurring in them, and we may suppose this electricity to flow like a current out by this wire, along any circuit formed by a conductor, and back to the battery again by this other wire. Such a current we may call a continuous current. Now let us see what effect a current of this kind has on a nerve connected with a muscle. We connect the wires of the battery with two platinum wires,

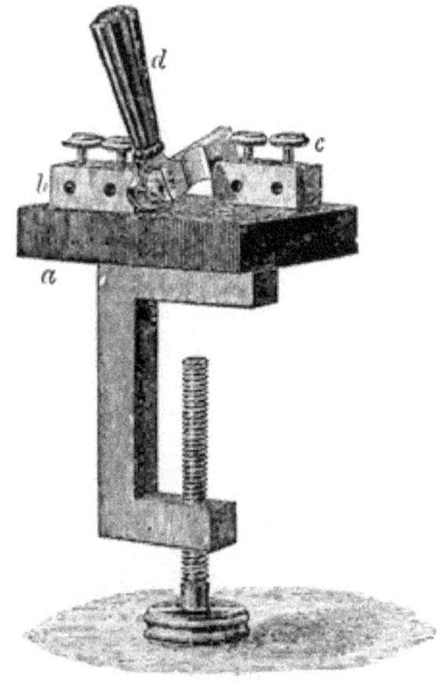

FIG. 17.—Key for opening or closing current.

over which the nerve has been stretched, and the muscle, as before, is connected with the muscle telegraph bell. To enable us to stop the passage of the current, or to send it on at pleasure, we place in the circuit what is called a key. It consists of a rectangular wooden frame, by which the instrument may be screwed to the table. On the top is a square block of vulcanite, *a*, bearing two rectangular bars of brass, *b* and *c*, which may be joined by the handle, *d*, carrying a horizontal piece of brass. Suppose wires from the battery are connected with *c* and *b*. The key is *closed* when the arm is horizontal, and the current runs along the horizontal piece of brass. On moving the handle, *d*, backwards and to the right, the brass arm is raised and the contact between *b* and *c* is broken. The key is then said to have been *opened* and the flow of the current is interrupted.

Well, you observe that when I close the key, and the current is sent through the nerve, the muscle gives a contraction, but it quickly relaxes, and no change, so far as the muscle is concerned, is visible while the current flows through the nerve. Now I open the key so as to stop the flow of the current through the nerve, and again there is a sudden twitch. If I open and shut the key quickly there is a twitch with each movement of opening and of shutting, and you see the muscle passing into the more permanent state of contraction that we called cramp or tetanus. It is apparently the suddenness with which the electric current enters the nerve and the suddenness with which it leaves it that irritates the nerve. The nerve is not irritated so as to cause contraction of the muscle *during the passage of the current through it.* Hence we would expect that currents or shocks of extremely short duration would be very irritating, and this is exactly what experiment proves. We obtain such almost instantaneous currents by the use of an instrument called an induction coil or inductorium.

To explain this to you let me show you a famous and far-reaching experiment first made by Faraday in the laboratory downstairs, by which he discovered the method of obtaining what has since been called Faradic electricity, or electricity by induction. Here is a galvanometer, an instrument used for detecting electric currents. It consists of a coil of wire, in the centre of which is a freely suspended magnetic needle, so hanging that the needle is in the same plane as the coil of wire. A small silvered mirror is attached to the needle, and you observe the mirror reflects upon this scale or screen a beam of light from a lamp placed in front of the galvanometer. A very feeble current passing round this coil deflects the needle, and the deflection is seen by the movement of the spot of light, either to one side or to the other, according to the direction of the movement of the needle. These two large bobbins of fine wire form our induction coil. You observe they are not connected.

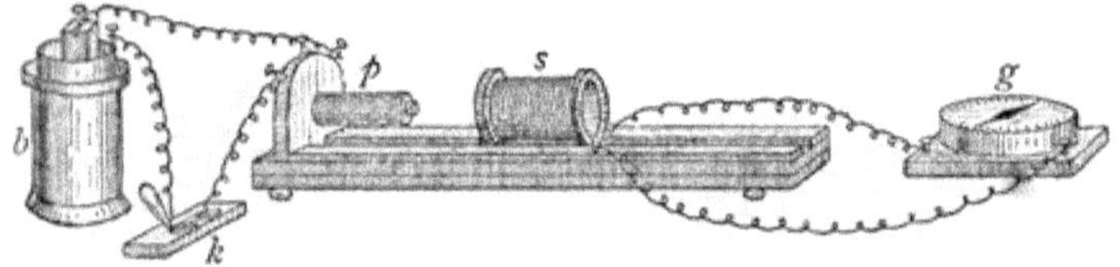

FIG. 18.—Arrangement of apparatus for demonstrating Faradic currents. *b*, galvanic element; *p* primary and *s* secondary coil; *g*, galvanometer. In the experiment a reflecting Thomson galvanometer was used.

In the circuit of the one we place a small battery and a key. In the circuit of the other we introduce the galvanometer. Watch the spot of light. When I close the key you observe an instantaneous movement of the spot of light. It swings to one side and then comes back, showing that the current passing through the galvanometer circuit is momentary. I now open the key, and you see again a momentary swing of the needle of the galvanometer, as indicated by the movement of the spot of light, but it is in the opposite direction. These instantaneous currents from what we call the

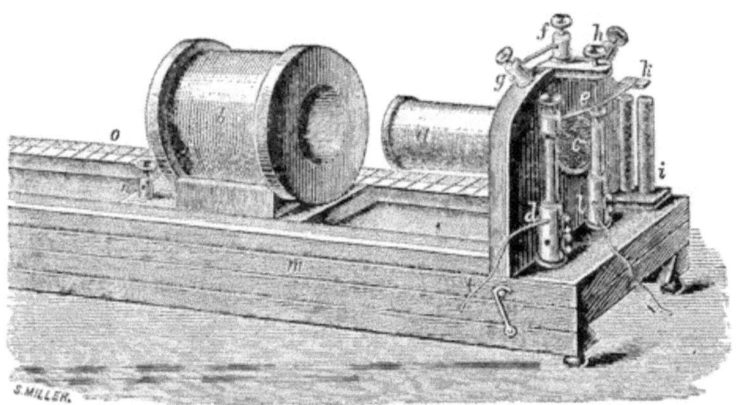

FIG. 19.—Induction coil of du Bois Reymond. *a*, primary coil; *b*, secondary coil; *c*, bunch of wires in centre of primary coil for increasing intensity of induction currents; *d*, binding screw for attachment of wire from galvanic element. The current passes up the pillar *d*, along steel spring to *e*, thence to the screw, the point of which touches the back of the spring at *e*; from *f* through wire of primary coil to *i*, round the two pillars of soft iron *i*, which it renders magnetic, and thus draws down the head of the spring *k*; this interrupts the current at *e*, thus breaking the contact of the spring at the screw point. When the current is thus interrupted, the spring flies up by its elasticity and again establishes the circuit at *e*. This interruption was originally invented by P. Wagner.

secondary coil are the Faradic or induced currents. They last only a minute fraction of a second of time. An induction coil, then, con sists of a primary coil, with which a galvanic cell is connected, and a secondary coil. When the current is stopped from flowing through the primary coil, that is on opening the key, the current induced in the secondary coil is in the *same* direction as that of the primary, but when the current is allowed to flow through the primary, as happens when the key is closed, the induced current travels in the *opposite* direction. If, then, we open and close the primary with great rapidity, say open one hundred times and close one hundred times per second, we obtain a short secondary shock with each opening and with each closing, or two hundred shocks per second. This rapid opening and closing is accomplished by a vibrating spring, which works automatically at the end of the instrument; and we graduate the strength of the shocks by increasing or diminishing the distance of the secondary coil from the primary, the shocks becoming weaker as we draw the secondary away from the primary.

The shocks from this instrument are much more irritating than those obtained from an ordinary battery. Thus, you see, I can hold the wires coming direct from the battery without being conscious of any irritation, but if I send this same current through the primary, I can hardly touch the wires coming from the secondary. The currents from the secondary are momentary in duration, and as they can be localised, they are used by physiologists as convenient stimuli for nerves and muscles.

Here is a large induction coil. You see the powerful discharges it gives, and when we send these through one of the late Mr. Warren de la Rue's vacuum tubes, containing a residue of carbonic acid, we get a magnificent luminous streak of quivering light in the tube, with beautiful transverse markings or bands.

A living being may be electrified positively or negatively and have no sensation caused by the electrification. You see here a frog sitting under this bell glass on a tin-plate connected with one of the dischargers of this large Wimshurst influence machine driven by an electric motor. The tin-plate and the frog's body are highly electrified, as you see by the sparks that fly out when I bring my finger near the tin-plate; but the frog is undisturbed so long as I do not touch it. We will not put it to the pain of having a tetanic spasm by touching it, but instead of the frog we will electrify my assistant standing on an insulated stool. So long as I abstain from touching him he feels nothing, but you see, if I touch his head, or his neck, or the tip of his nose, how the sparks fly out, and he then feels a smart and disagreeable sensation. During electrification we feel nothing, but it is only when we pass from one stage of electrification to another that we have a sensation, and the more rapidly this change takes place, the more irritating the sensation is.

We have now seen how nerves and muscles may be irritated in a definite and precise way, and we have found that the irritation of its nerve causes a contraction of a muscle. In the case of a single twitch, however, the movement is too rapid to be appreciated, and still less analysed, by the unaided eye. We cannot tell, for example, whether the contraction occurs in a shorter time than the relaxation, and still less whether the contraction is at a uniform rate in time, or whether it contracts faster at the beginning and more slowly towards the end, or the reverse. We must not, therefore, trust only to our senses in the study of rapid movements, and we call to our aid what is termed the graphic method of registering movement.

Suppose I hold my pen between my thumb and first two fingers in the ordinary way, and extend the thumb and fingers so as to move the pen upwards on the paper. I produce a line thus— The length of the line from *a* to *b* will show the amount of movement of the pen point, but it will not tell me anything about the rate of movement, nor whether the pen moved with uniform velocity in passing from *a* to *b*. Suppose now that I repeat the experiment of drawing the line from *a* to *b*; but, on this occasion, suppose the paper is moving with uniform velocity from right to left while I draw the line. It will then be found that I describe a curve something like this—

b

a

Fig. 20.

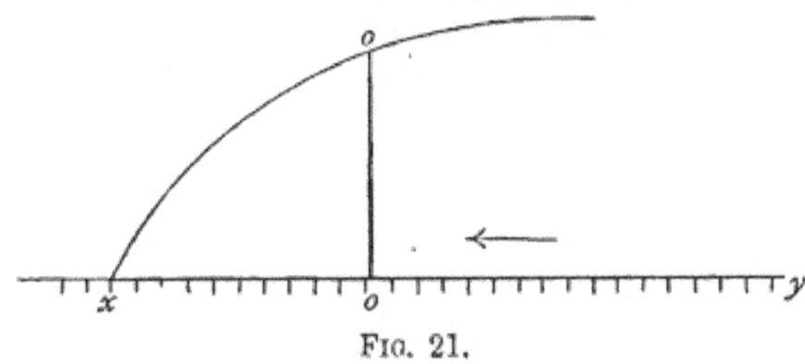

Fig. 21.

The paper was moving in the direction of

the arrow ←. If now I draw a line $x\ y$, and divide it into equal parts, representing equal periods of time, say tenths of a second, and if I drop a perpendicular line from the curve down to $x\ y$, say from o to o, I will see at once where my pen point was at that instant of time. Suppose, again, I extended my fingers and then relaxed them so as to draw a line on the paper at rest, I would again draw a line like $a\ b$; but if I repeated the experiment, with the paper moving quickly from right to left, I would describe a curve thus—

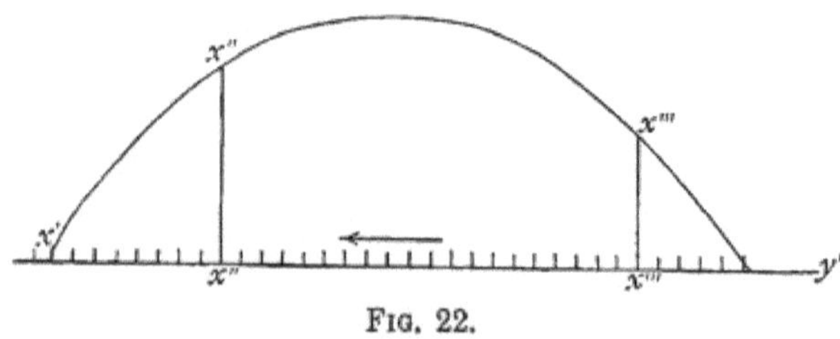

Fig. 22.

and if time were indicated in tenths of a second by a line $x'\ y'$, the position of the pen point at any instant between x' and y' would be found by dropping perpendiculars, as from x'' to x'' or from x''' to x'''. If my pen point travelled faster in going up than in coming down, supposing the paper moved with the same velocity as in the previous experiment, the curve would be something like this—

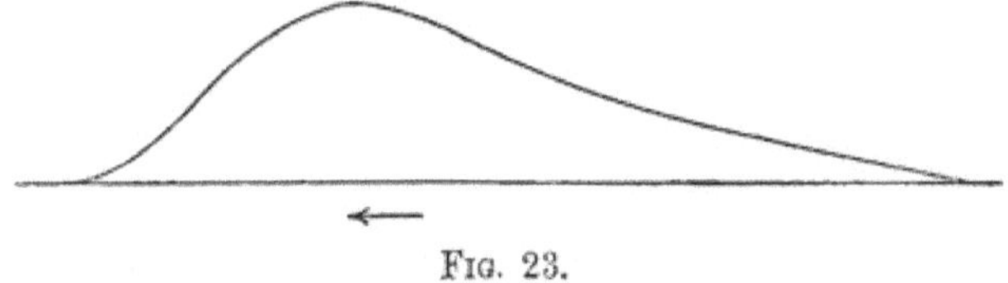

Fig. 23.

and if it travelled slower in going up than in coming down it would vary its form to—

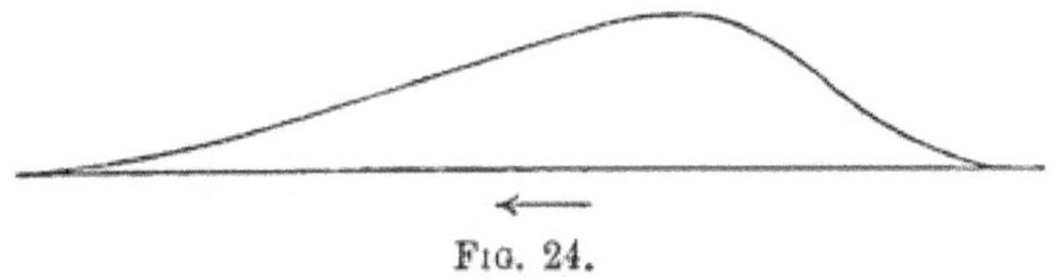

Fig. 24.

Thus by recording rapid movements on a quickly moving surface, such as the surface of a drum or cylinder, or on a glass plate travelling horizontally, we get information regarding phases or variations of movement which we could not otherwise obtain. This illustrates the essential principle of the graphic method, a method of great value in all sciences dealing with movement, and not least to physiological science. It is not a modern method, although in later times its use has been enormously extended. In 1734, the Marquis d'Ons-en-Bray described an anemometer, an instrument for recording the velocity of the wind, which registered its movements on a sheet of paper rolled round a cylinder moved by clockwork. Magellan, in 1779, made designs for an instrument for recording automatically many meteorological phenomena. In 1794, Rutherford constructed a thermometer by which curves of changing temperatures were marked on blackened paper. Thomas Young, one of the founders of this Institution, in 1800, showed how time could be measured on the surface of a cylinder moving at a uniform speed. The celebrated James Watt devised a method of tracing the movements of the indicator of his engine on a cylinder rotated by the engine itself. Thus he obtained a curve showing variations of steam-pressure at different times. During the past thirty years, numerous ingenious instruments have been invented by physiologists for recording movements; but no physiologist has done so much in this direction as Professor Marey of the College du France, to whom we are largely indebted for the development of the graphic method to its present condition of precision and convenience.

Our arrangements for studying muscle are not yet complete. We have already seen that a muscular contraction is so rapid as to make it impossible to follow all its phases with the unaided eye. The questions at once occur to us: how long a time does it take to contract? if it does not contract at the same rate throughout its contraction, for how long a time does it contract quickly, and for how long a time does it contract more slowly? These questions lead us at once into another department of inquiry of immense importance in science—the measurement of time, and more especially the measurement of minute periods of time. We all recognise more or less the value of time, and the busier we are the more we value what we call fragments of time. On a long summer day in the holidays, when we have not much to do except gratefully to enjoy the beauties of nature and the sense of physical well-being, a quarter of an hour, or even an hour, is not appreciated as of much value; but when we have a great deal to do, as in winter, when every moment seems to be occupied, five or ten minutes are felt to be precious. We can put a good deal into ten minutes when we are very busy. Science teaches us the value of short periods of time, because nature, which science deals with, is always busy, filling up every moment with some kind of work. We poor mortals count time according to our own wants, and we think a second is about as short a time as we need pay any attention to; but nature does work in much shorter periods of time than a second. A lightning flash occupies only the millionth part of a second, and a dragon-fly's wing, as it passes through the air on a summer day, is quivering many hundred times a second. The fact is, our appreciation of intervals of time by the senses is very limited, and when events happen that are not more than the tenth of a second apart, we are apt to think they are simultaneous. The reason of this is that we need time to think, or, in other words, time is occupied by the processes that go on in our brains. Time is needed for perception, and if two events follow each other so rapidly that during the time we are perceiving the first the second comes on, we blur the one perception into the other, and we fail to notice the interval of time between them. And yet during that interval much may have happened outside of us in connection with each event, of which, however, we are unconscious.

Science, then, demands the accurate measurement of time, and no small fraction of time is too insignificant to be of importance. The twentieth, the hundredth, the thousandth of a second is to science as precious as an hour may be to many people. It may be said without irreverence that to science a day is as a thousand years, and a thousand years as one day. The methods of science for this purpose are founded on two sets of appliances : first, the use of instruments, like a tuning-fork, that are found to vibrate or move at a known rate, say one hundred or five hundred times per second ; and, second, the application of the graphic method, by causing these instruments to record their movements on a rapidly moving surface, such as that of a cylinder travelling with great velocity. All such appliances are called chronographs, or time-writers. Let me illustrate to you the use of several of these ingenious appliances.

Here is a ball at the end of a long string suspended from the roof of the theatre. It constitutes a pendulum. During its entire swing it occupies three seconds. Now observe I can amplify the distance of the swing at pleasure, but the time in which the ball travels this distance is always the same. Suppose I make it swing through a distance of twelve feet, then (making allowance for the law that regulates the movements of a pendulum) four feet would roughly represent one second, and one inch would represent the one-forty-eighth of a second. Thus by amplifying the extent of the swing and subdividing we get a notion of small periods of time, such as the one-hundredth or the one-thousandth of a second. If I shorten the length of the string of the pendulum, the ball swings faster, but through a shorter distance in each swing. Carry this on until the ball, suppose it now to be very small, moves, say a hundred times backwards and forwards in a second, and you pass on in thought to the delicate instruments we shall now consider.

The first instrument, devised by Thomas Young, is the revolving cylinder. Suppose this cylinder revolved at a uniform rate by means of clockwork. Suppose the surface of the cylinder to be divided by eight lines, drawn parallel to its axis at equal distances from each other, and that the cylinder makes one revolu-

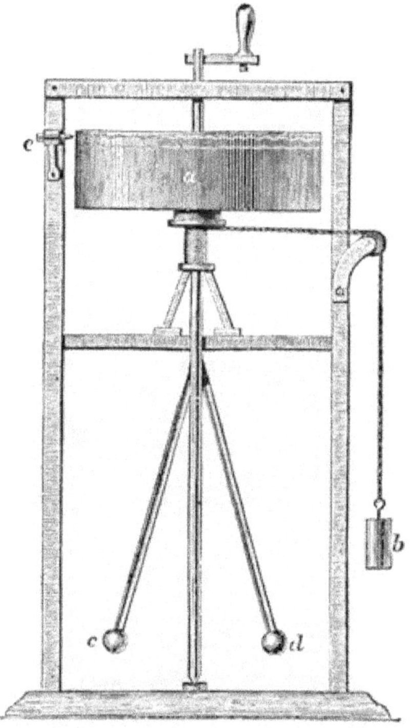

FIG. 25.—Original chronograph of Thomas Young. *a*, cylinder rotating on vertical axis; *b*, falling weight acting as motive power; *c d*, small balls for regulating by centrifugal action the velocity of the cylinder; *e*, marker recording a line on the cylinder.

tion in one second, the distance between two of the lines will represent the one-eighth of a second, and any movement drawing a curve on the cylinder between the two lines must have happened during that interval of time. It is evident that, by drawing the vertical lines closer to each other, much shorter intervals, even to the one-thousandth of a second, may be measured with accuracy, provided the cylinder moves with uniform velocity. It is not easy to secure the latter condition. You see when I start this cylinder it moves slowly at first, and gathers speed as it goes on, and even when it

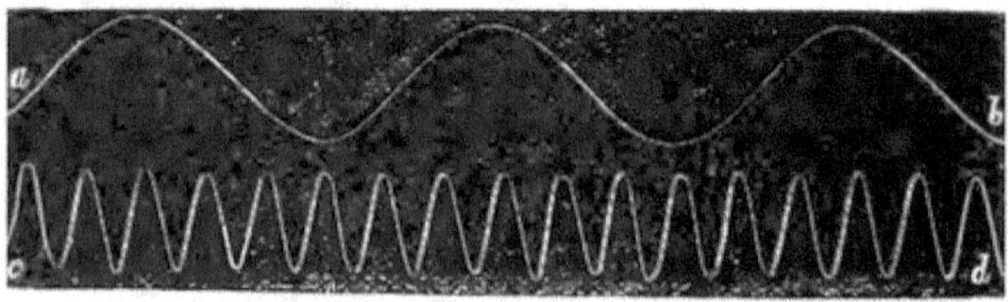

FIG. 26.—Tracings of the vibrations of a tuning-fork, ten vibrations per second. *a b*, cylinder moving rapidly; *c d*, cylinder moving slowly.

attains full speed I have no guarantee that it is then travelling at a uniform rate. It may make short spurts, or, as the spring becomes unwound, it may by and by move more slowly. The method by the cylinder, therefore, is not sufficiently accurate.

Thomas Young was also the first to devise the method of inscribing on a rotating cylinder the vibrations of a rod bearing a very light style or marker. These describe undulations on the cylinder, and the undulations correspond to equal periods of time. No instruments vibrate with greater uniformity than tuning-forks. Duhamel was the first to apply to one of the limbs of a tuning-fork a small marker, and to bring this marker against a rapidly moving surface, like the surface of this blackened cylinder. Undulations or waves are thus described. The more rapid the movement of the cylinder, the longer will be the waves, as you see in the experiment, and as are represented in this diagram (Fig. 26).

It is difficult to apply the vibrating limb of a tuning-fork to a cylinder, more especially if other recording apparatus is adjusted to the cylinder at the same time. Suppose we use the fork simply for interrupting the current, and this it will do with great regularity, we might interpolate in the circuit a little electromagnetic appliance, having a keeper, to which a marker is attached. This is the apparatus you see here. It consists of a battery, an interrupting tuning-fork, and an electromagnetic instrument called the chronograph or marker. The chronograph consists of a fine marker, *c*, fixed to the end of a steel spring, and armed with a mass of steel, somewhat wedge-shaped, which fits in between two small keepers,

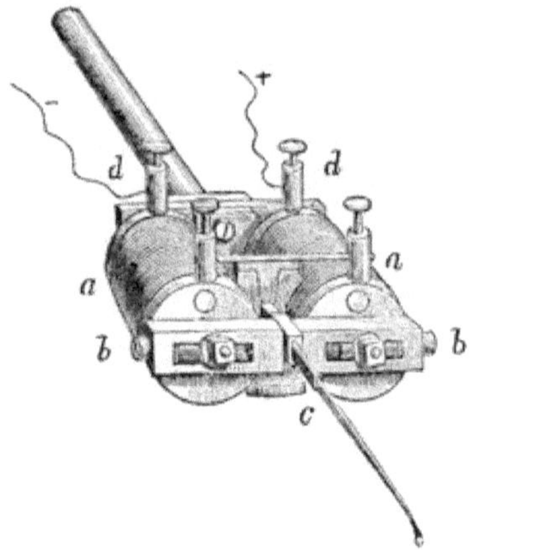

Fig. 27.—Electric chronograph. For description see text.

b b, of the electro-magnets, *a a*. The tuning-fork interrupts the current from the battery.

Fig. 28.—Interrupting or chronographic tuning-fork.

This it does automatically. When the iron of the electro-magnet between the limbs of the tuning-fork becomes magnetic, the limbs are drawn together, and a small bit of platinum wire fixed to one of them is removed from contact with a platinum surface, so as to break the circuit. On the circuit being thus broken, the electro-magnet ceases to act; the limbs of the fork, by their elasticity, spring back to their original position; and thus the platinum wire is again brought into contact with the platinum surface. Thus again the circuit is completed and the action is repeated. In this way, the marker of the chronograph vibrates in unison with the fork, and you see, when I bring it into contact with the cylinder, a beautiful series of little waves is described, each little wave representing the one-hundreth of a second. Let us take a rough illustration to show the value of recording time in this way. Suppose I draw my hand, bearing a pencil, quickly from left to right and then back again. We wish to know the time occupied by that movement. I set the cylinder in rapid motion, the time is registered by the chronograph, and I now make the wished-for movement, bringing my pencil point into contact with the blackened surface. Here is the result. We have only to count the number of little waves made by the chronograph corresponding to the big wave made by my hand. You see we have twenty little waves. Thus twenty hundredths, or one-fifth of a second, represents the time occupied by the rapid movement of my hand.

It is impossible, in a lecture-theatre like this, to show to a large audience, such as I have the honour to address, the curves described by any recording apparatus on a rotating drum or cylinder. What we need is an arrangement by which the curves can be at once projected on a screen by the electric light. Indicating my wants to Mr. Horace Darwin, of the Cambridge Scientific Instrument Company, we have been able to devise the apparatus now before you, and which we will call the R. I. Railway. As it will be used chiefly for obtaining the curves of contracting muscles, it may well be named the railway myograph. You see it is a triangular frame, carrying a large plate of glass secured in a vertical position. The glass is blackened with soot in a smoky flame, and, of course, the soot prevents any light from passing through it; but if any soot is rubbed off, as when we draw a line on the plate with the point of the pencil, the light shines through,

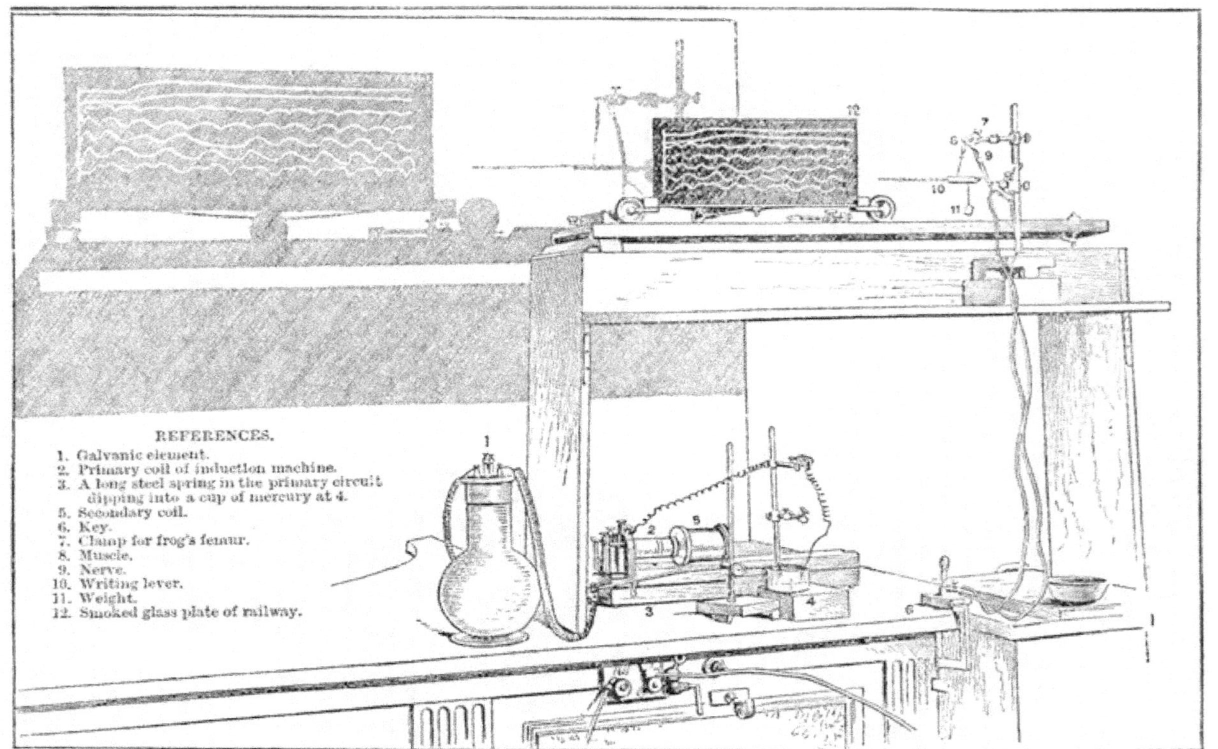

FIG. 29.—The new railway myograph, as arranged on the lecture-table at the Royal Institution. The shadow of the apparatus was thrown on the screen. The apparatus is depicted as it was arranged for the study of tetanus.

and the line appears as a bright shiny line on the screen. The wheeled car bearing the plate of glass is drawn up to the end of the long board, and this you observe is not level, but we may incline it by turning this screw which raises one end. A catch holds the car in position, and when the car is in position a strong spring is put on the stretch. When the catch is released the car runs down the incline, and at the same time the spring recoils, and, pulling on a lever, sends the car along with great velocity to the other end of the board. As it runs along we have an arrangement by which, when about midway in its course, the car breaks an electric circuit. We shall not use the "break" in the present experiment.

Now we shall show you a record of the movements of our chronograph worked by this tuning-fork, which is vibrating one hundred times per second. Mr. Brodie, you observe, brings the car up and fixes the catch. The spring is on the stretch. He now adjusts the marker of the chronograph on the glass plate and you hear the humming of the fork. All being ready, Mr. Brodie releases the catch, the car, carrying the glass plate, dashes across, runs into the electric beam, and is caught firmly by a spring that prevents it from rebounding. Mr. Heath had previously adjusted the electric lantern, and you now see on the screen the beautiful sinuous line, each wave of which represents the one-hundredth of a second.

You will be thinking that all this has not much to do with muscle, and I fear the description of these appliances may have been a little wearisome to you. But we must know something about the methods by which we attain results in science. This gives one a better appreciation, a better grasp, as I may say, of these results, and it shows us how men have got over difficulties in their attempts to explore phenomena. The recognition of how they have done this is an education by itself.

Now come to another physiological experiment. I have fitted up on this stand a number of pieces of apparatus, all intended for studying muscular contraction. First, at the top, you see a brass forceps which tightly holds the thigh bone of the frog's leg. You see again the gastrocnemius muscle hanging down, with its tendon, by means of a hook, fixed to this very light lever. Underneath the lever I have suspended a weight weighing 10 grammes, that is, about 150 grains, about the third of an ounce, which the muscle will be obliged to lift when it contracts. Then you see the sciatic nerve is stretched over these platinum wires, which we call electrodes, and the wires come from the secondary coil of our induction

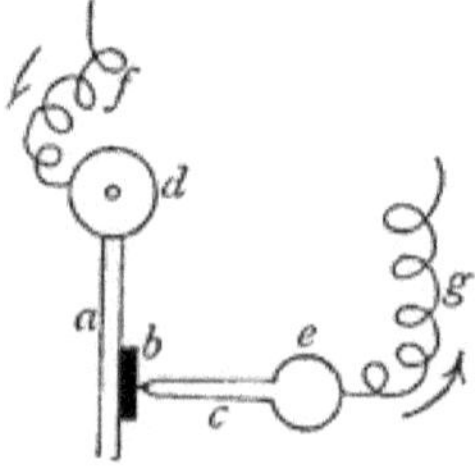

FIG. 30.—Diagram showing the break in the railway myograph. Current enters at *d*, passes along arm of brass *a*, having a bit of platinum at *b*, thence through point of screw *c* to *e*, and back to battery by *g*. When the plate of the railway runs across, it knocks *a* aside and opens current at *b*.

machine. Here is the primary coil of the induction machine. I have interposed in its circuit this galvanic element or cell, and we will allow our railway myograph to break the circuit of the primary coil as it rushes down the railway incline. Thus when the break is opened, the opening shock will go from the secondary coil to the nerve. Now, the moment the break is opened will practically coincide with the moment the nerve receives the shock, because the electricity travels with such enormous rapidity in the coils and along the wires to the nerve as to make the time between opening the break and irritating the nerve practically nothing. If I knew the moment the nerve was stimulated, it would be interesting to ascertain if the muscle contracted at that moment. Now we can easily record this moment by first of all causing the muscle to contract by bringing the car slowly up to the break till it opens it. When the muscle contracts, it makes a mark, which indicates the moment the break will be opened when we perform the actual experiment. At that moment the nerve is irritated, and if the muscle contracts at the same instant, the beginning of its upward curve should exactly coincide with the mark of the signal. Now we shall perform the experiment. Mr. Brodie has got the railway ready for starting. He opens the break, releases the catch, and lets the glass plate dash onwards. The muscle, of course, does not contract, as the nerve has not received a shock, and we have only a horizontal line drawn thus—

Then Mr. Brodie, in the next place, closes the break, brings the carriage slowly up to it, and opens the break gently. The moment this is done, a shock (the opening shock) comes from the secondary coil and the muscle makes a mark *a b* thus—

He then again closes the break, brings up the railway to the catch, releases the catch, the railway dashes across, opens the break, the nerve gets the shock from the secondary coil, and the muscle contracts. But you observe the muscle has contracted a little later than the instant the break was opened, so that we get this curve (Fig. 31).

You notice the momentary twitch. Here is the tracing. You see the muscle curve has begun a little later than the mark made by the signal, that is to say, the muscle did not contract the instant the nerve got the shock. A little time intervened, something like the one-hundredth of a second, during which nothing visible happened, and then the muscle contracted. This period is called the period of latent stimulation. In the experiment we have just made, it does not

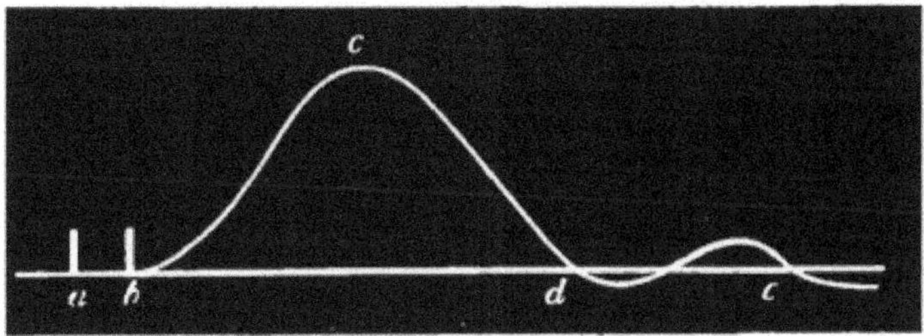

FIG. 31.—Curve of a single muscular twitch as taken with the railway myograph. *a* to *b*, period of latent stimulation; *b* to *c*, contraction; and *c* to *d*, relaxation of muscle; *e*, small secondary wave, probably due to movement of lever.

exactly represent the period during which the muscle rested after receiving the molecular disturbance of the nerve-current, as I have already explained, and we must deduct from it the time occupied by the nerve-current travelling down the little bit of nerve. However, it is interesting to know that there is a short period during which changes are probably happening in the muscle before it contracts. In the latent period, the muscle is preparing for making the movement. Very refined methods show that the latent period is shorter than the one-hundredth part of a second. Professor

Burdon Sanderson has found, by a photographic method, not so liable to experimental errors as the one I have shown you, that in the muscles of the frog it is about the one-two-hundredth part of a second. Probably it is even shorter in the muscles of the higher animals. Their muscular substance is more unstable than that of a frog, and it goes off more rapidly under the nervous stimulus. Research also shows that probably in all living matter submitted to a stimulus there is a latent period, a period in which molecular changes are happening which precede, and possibly end in, the particular phenomenon manifested by the living matter. Thus when the nervous stimulus reaches the cell of a secreting gland, or a blood-vessel, or a nerve cell in the spinal cord or brain, it does not produce an immediate effect, but excites changes which occupy time.

When we next meet we shall study tetanus or cramp, and how a nerve probably acts on a muscle.

LECTURE III

Study of tetanus or cramp—Simple contraction and muscle curve—Muscle sound—Elasticity of muscle—Inherent irritability of muscle—White blood corpuscles—Cilia—Structure of muscle—End-plates in muscle.

To-DAY we shall, in the first place, study tetanus or cramp. For this purpose, I have fitted up the usual preparation, with which

Fig. 32.—Steel spring for making and breaking the primary circuit of induction coil. The point on the left is caused to dip into mercury, and thus make contact. See Fig. 29.

you are now familiar, and I have made arrangements for opening and closing the primary circuit of the induction machine by means of a long flat spring. Here is the battery. Let us follow the wire leading the current from the positive pole of the battery to the primary coil. Then the current passes through the coil and out by this wire to a little cup containing mercury. When the needle on the under surface of the end of the spring dips into the mercury, the current will pass through the mercury, along the spring, and then, by this other wire, back to the battery, entering it at the negative pole. I have also interpolated an electro-magnetic signal in the circuit, so that each vibration of the spring is also registered on the plate of the railway myograph.

Let us begin by sending a very few shocks per second to the nerve. This we can do by using the spring at its full length, thus causing it to vibrate slowly. You see the muscle twitching each time the spring dips into the mercury and comes out again, and we get a curve in which the separate contractions can be seen (see Fig. 33, *b*, *c*, *d*, *e*). But already, you will observe, the contractions are, as it were, piled upon each other, *f*. Thus the muscle contracts with the first shock; then it begins to relax; but before it has had time to relax it again receives a shock, and it again contracts. Then it again relaxes; but it does not relax so much as before, ere it receives the third shock, and so on. Thus we get a notched curve, or a curve something like a staircase, in which the

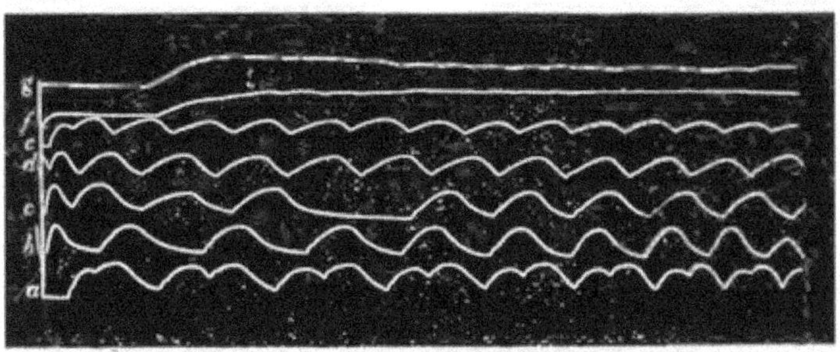

FIG. 33.—Curves showing the production of tetanus, as taken in the rapidly moving plate of the railway myograph. Observe in all the experiments from *a* to *e* the muscle had time to relax between the shocks; at *f* tetanus began to appear, and in *g* it was complete. The curves here shown are one-fifth of their real size.

successive steps become smaller and smaller as we ascend (see Figs. 34, 35, and 36).

By shortening the spring, we quicken the period of its vibration, and thus we increase the number of shocks per second. Now you observe the individual contractions are smaller and closer together; but if we look at the tracing carefully, we find the same stair-like character of the curve, only the steps are smaller (Fig. 35). Again, still further shorten the spring and increase the number of shocks. The lever is at once pulled up as far as it will go; but it quivers with each shock, and the curve shows a number of little teeth along its summit (Fig. 36). Make the spring still shorter, and you find the quivering disappears, and the uniform curve of tetanus shows itself,

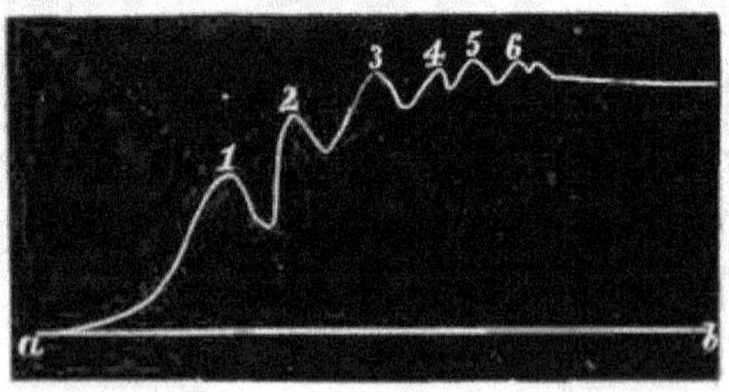

FIG. 34.—Tracing of a muscle passing into a tetanic state. The first shock was transmitted to the nerve at *a*, the second an instant after 1, the third an instant after 2, and so on. It will be observed that with each succeeding shock the muscle becomes shorter, though the amount of shortening with each shock is less.

a curve having a long flat summit presenting no teeth (Fig. 37).

The diagrams in Figs. 34 to 37 show curves taken on a slowly moving drum.

This experiment demonstrates that tetanus is produced by a fusion or adding together of small contractions. One shock causes one contraction; two shocks, closely following, cause two contractions so far blending into each other; three shocks, closely following, cause three contractions, still further blending, and so on, until the shocks come so fast that the individual contractions are all fused together to form a curve. Tetanus, therefore, is not one contraction, but a state brought about by the

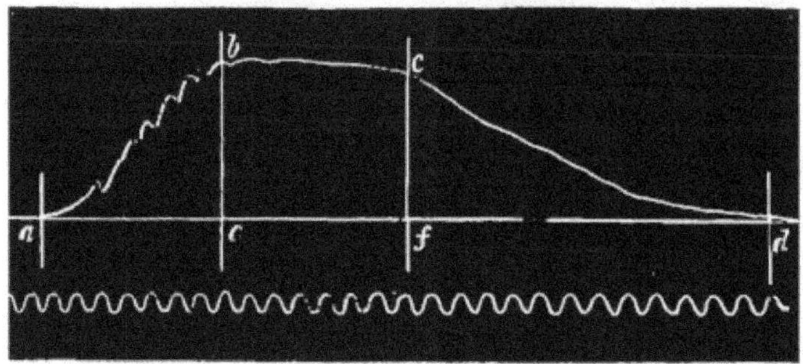

FIG. 35.—Curve showing the production of tetanus. *a* to *b*, individual contractions; *b* to *c*, muscle now tetanic. The slope of line from *b* to *c* shows that muscle is becoming fatigued; *e*, indicates moment when induction shocks stopped; *c d*, slow relaxation.

fusion or summation of many contractions. With the frog's muscle, about fifteen shocks per second are sufficient to cause tetanus; the muscles of a tortoise require only two to three; the muscles of a rabbit from ten to twenty; the muscles of birds about seventy; and the muscles of insects over three hundred per second. If the number of shocks is very much increased, even to as many as twenty-four thousand per second, tetanus is still produced.

Let us now take a tracing of a simple muscular contraction on a blackened surface moving with great rapidity. This is usually

FIG. 36.—Curve showing production of tetanus. *a* to *b*, result of first shock; then observe the cumulative or gathering-up effect of the successive shocks as shown by gradual ascent of line from *b* to *e*; stoppage at *e* of shocks; *e f*, gradual relaxation.

done by causing the muscle to record its movement on a glass plate forming the bob of

FIG. 37.—Tetanus curve produced by numerous hocks from induction coil. The individual contractions are no longer seen.

a seconds pendulum which is allowed to make only one swing, and the tracing obtained is shown in this diagram. I found it was not easy to fit up such a pendulum in this lecture-room, and we shall therefore use the railway myograph. Mr. Brodie will cause the myograph to interrupt the primary circuit of the coil only once, and thus send only *one* shock to the nerve. The muscle will contract and describe its curve. Here it is. The curve is so long as to allow us to analyse it carefully. We

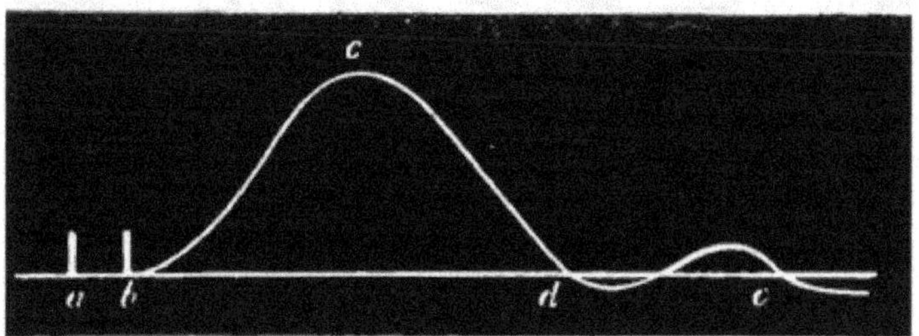

Fig. 33.—Curve of singular muscular twitch as taken by railway myograph.

have first the latent period already discussed. This occupies say one-two-hundredth of a second. This is followed by the stage of contraction, from the moment the muscle begins to shorten to the time when it reaches its greatest degree of contraction. In this stage, as you see by the varying slopes of different parts of the curve, the muscle usually contracts, at first slowly, then rapidly, and again more slowly, and the time occupied is three-hundredths to four-hundredths of a second. Next, the muscle at once begins to relax, at first slowly, then rapidly, and again more slowly, and the time, shorter than the time of contraction, is less than three-hundredths of a second. Lastly, as a rule, we find a few smaller waves, as if, in recovering itself, the muscle had been thrown into a kind of vibration. Sometimes the muscle, unless it be drawn out by a weight attached to it, does not return at once to its original length, but remains somewhat shortened. This occurs readily in muscles that are fatigued.

The question at once suggests itself as to whether the contractions of the muscles in our bodies are of the nature of twitches, that is, single contractions, or of tetanus. We can apparently flex and extend the arm with great rapidity, and one would naturally suppose that such contractions of the biceps muscle in front, and of the triceps muscle behind, were simple contractions. There are strong grounds for holding, however, that they are not so, but that the movement is really a short tetanic contraction. There can be no doubt that when we firmly contract a muscle and maintain the contraction, the muscle is in a state of tetanus, in which the quivers of the partial contractions can be seen. We all know how difficult it is to keep the hand or arm quite steady. They tremble, and we can feel the vibration. This can be demonstrated.

I have here a strong spring placed horizontally, and while it is firmly secured at one end I can pull upon the other. To show its vibrations, we have attached to the spring, and at right angles to it, a thin rod, the upper end of which is in the electric beam, and you see its shadow on the wall. When I pull as strongly as I can, you see the upper end of the rod vibrating. The muscles of my arm are in a state of strong contraction, of physiological tetanus, but they cannot remain permanently contracted. At one instant they relax a little, and the elastic recoil of the spring stretches them; then they contract for a moment, and so on. Thus they are vibrating.

While I was arranging this experiment I considered how I might be able to show you the time occupied by voluntary movements, say those of the fingers in writing a letter. It can be done in this way. Here is a tuning-fork worked by an electro-magnet, and vibrating about 240 times per second. I have an arrangement by which I can attach a glass plate to the side of one of the limbs of the fork, and I have placed exact counterpoises on the other limb. I now blacken the slip of glass

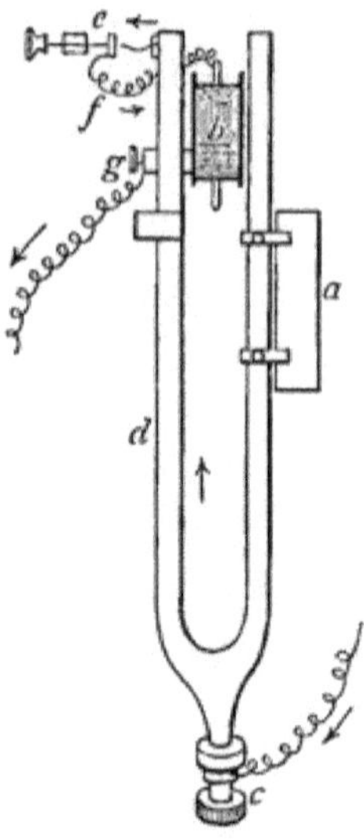

Fig. 39.—Tuning-fork arranged for causing a smoked-glass plate, *a*, to vibrate. *b*, electro-magnet; current passes in at *c*, along limb of fork, *d*, along platinum wire to platinum contact at *e*, thence by wire, *f*, to electro-magnet, *b*, thence back to battery by wire from *g*. When *b* is magnetised the limbs of the fork are drawn together and contact is broken at *e*. The limbs then fly back and again make contact at *e*, and so on.

(an ordinary microscopic slide) in the smoky flame of this lamp and attach it to the fork. Set the fork vibrating. I shall now write something on the smoked surface with a needle point, taking care to write at right angles to the movements of the fork, and then Mr. Heath will place it in the lantern. Then you see the words, and you will observe that the letters show little curves, each of which represents

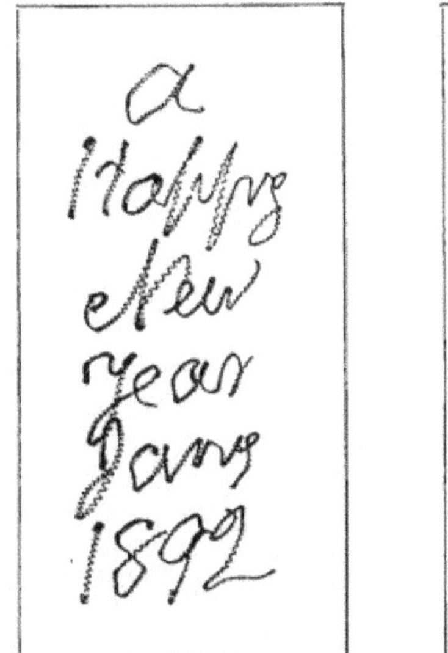

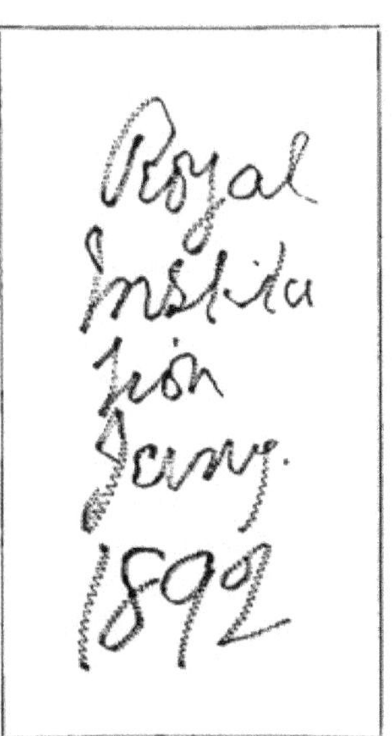

FIG. 40.—Time of making voluntary movements in writing letters. Each little wave is the one-two-hundred-and-fortieth of a second. Examine with a magnifying glass. In the experiment the lines were white on a black ground.

about one-two-hundred-and-fortieth of a second. By counting the little curves in any letter you can measure the time I occupied in writing it.

But let us return to the question of whether ordinary voluntary contractions are twitches or tetanic spasms. Another strong bit of evidence in favour of the occurrence of what we call physiological tetanus we find in the phenomenon of the so-called muscle sound. This was first observed by the celebrated Dr. Wollaston, a prominent man of science of his day, and one of those who took a deep interest in this Institution in its earlier years. His bust is on the staircase. He discovered that when a muscle contracts, and is maintained in a state of tension, it gives out a sound or tone. We can hear it by placing the ear over a muscle, like the biceps of a muscular person; or, in the dead of night, when all is still, by strongly pressing the teeth against each other by clenching the muscles of the jaws. You may hear it, I believe, simply by putting the tips of the index fingers into the ears and then contracting the muscles of the arms. Now you are aware that the pitch of a tone is determined by the number of vibrations per second made by the body that vibrates and gives out the tone. This tuning-fork, for example, vibrates 128 times per second and gives out a tone of low pitch, while this other one vibrates 8 times as fast, or 1004 vibrations per second, and consequently gives a tone of much higher pitch. The pitch of the muscular tone indicates that it is produced by about 19·5 vibrations per second. If so, it follows that in a persistent muscular contraction in a healthy person, the muscle must be vibrating or quivering that number of times per second; or, in other words, the contraction is a kind of tetanus produced by about twenty shocks per second. But the stimulus that causes a voluntary contraction comes from the nervous system, the impulse originally starting from the brain. This stimulus passes along the nerves to the muscles and is their normal stimulus; but these considerations show us that the nervous stimulus, whatever it may be, is not like a continuous current flowing from the brain in the nerves to the muscles, but that it is intermittent and is more comparable to a series of shocks sent out at the rate of from ten (as some hold) to about twenty per second. Thus you see the study of tetanus lets us get a glimpse of what is probably occurring in every voluntary movement.

We have seen that a muscle is contractile. Has it any other special properties? Here are two muscles hanging side by side of about equal size and equal weight. Each has a strong silk thread tied round its tendon and a hook attached to the thread. One of these muscles is fresh and the other is dead, and has been so for many hours. I shall suspend equal weights, so as to see how much each muscle will carry, and we will find that the dead muscle will tear sooner than the living one. Its cohesion is not so great. Further, we may notice that the dead muscle scarcely stretches when I put the weights on, but the living one stretches considerably, that is to say, the living muscle is extensible. Now watch the living one closely. I shall connect it with this lever, so that we may see its movements better. You observe when I put a weight on it, it stretches, and when I remove the weight it returns to almost, but not quite, its original length; that is to say, it is extensile and retractile. This property of becoming extended and then returning to its original length is still by most physiologists described under the name elasticity, but I prefer to retain this term for the property a muscle has of returning to its original length after it has contracted. If, however, we regard muscle as a passive structure, then we find that it is a slightly but perfectly elastic body. It yields to a weight, but when the weight is removed the muscle readily returns to its former length.

Let us examine more closely the behaviour of the muscle when it is extended by gradually increasing weights, and let us compare it with this band of india-rubber. We will fix the two, side by side, against a board on which we can mark the amount of extension in each case. Notice that with gradually increasing weights the india-rubber extends so that the amount of extension is directly proportional to the weight; that is to say, for each equal increment, the india-rubber is extended to an equal amount. Compare this with the muscle. You observe the first weight stretches it so much; if the weight is now doubled we do not get twice the amount of extension, but less than half; if the weight is again increased by one third, we get only a little more stretching; and so on, each increase on stretching becoming less and less. Thus, with the india-rubber, if we note the amount of stretching on the board by a number of vertical lines, each line representing a uniform increase in the weight, and if we join the end of these lines we get a straight line. On the other hand, by performing the same experiment on a muscle, the line joining the vertical lines that represent the stretchings is not straight but it is a curve, the curve of a hyperbola, in mathematical language; that is to say, a curve which constantly tends to become parallel with a horizontal line in the same plane but never reaches this condition.

The fact that muscles can be stretched, and that they return to their first length when the stretching force ceases to act, is of great importance to their mode of action. In the body, the muscles are always partially on the stretch. They are never "on the slack," to use a familiar phrase, and they are ready to exert a pull on the bones to which they are attached, the instant they begin to contract. Thus no power is wasted. Further, as has been ingeniously shown by Professor Marey, the elasticity breaks the force of the shock produced by the sudden contraction of the muscle, and the energy of the contraction is expended more gradually and effectively than if the muscle had been non-elastic. Suppose a horse drawing a cart along a rough road by non-elastic ropes or chains attached to its harness. It would receive, as it exerted its power, a number of jars and jolts which would not only be unpleasant, but which would absorb and dissipate some of the energy it was expending. The animal would be more comfortable, and it would work more efficiently, if a number of elastic structures were interposed between it and the cart. The jolts would be taken up by the elastic structures, and the horse, instead of having to pull in a spasmodic way, would pull steadily and without jerks. The amount of effective work would thus be increased.

I am inclined, however, to think that the elasticity of muscle plays even a more important rôle than this, as has recently been ably advocated by a great French physiologist, Professor Chauveau. It has long been known that a contracted muscle can be stretched to a given extent by a smaller weight than is required to stretch it when it is at rest and not contracted. This has led physiologists to say that by contraction the elasticity of the muscle is diminished in amount while it is still perfect; that is to say, a contracted muscle is easily stretched, and when the stretching force is removed, it returns at once to its former length. But may we not say that the contractility develops an elastic force different from the mere extensibility and retractility we have already studied, and that this elastic force does the work of producing movement? A contracted muscle, according to this view, is like a strong band of india-rubber pulling on two pieces of wood and drawing them together. In like manner, when a muscle contracts, a similar elastic power is developed that pulls upon the bones and causes movement. When the contraction ceases, the muscle returns to its original length, again by elasticity, but acting in the opposite way.

What is this property of irritability by virtue of which the muscle responds to a stimulus and contracts? Is it something possessed by muscles alone, or do we find it anywhere else? Upon this point, at one time, there was a keen controversy. In 1760, a great Swiss physiologist, Haller, wrote upon muscular irritability. Before this date, it was commonly held that the property of irritability was derived from the nerves, and it was supposed that the nerves conveyed the power or force manifested by the muscles. He found, however, that muscles remained contractile after their nerves had been divided, and even after the muscles had been removed from the body. He also observed contractility in certain plants destitute of nerves, and in some of the lower animals, in the bodies of which no nervous structures had been found. He arrived, therefore, at the conclusion that the property was inherent in muscular fibre itself, a *vis insita.* His views were strongly opposed by Robert Whytt, a professor in the University of Edinburgh, and the progenitor of the famous novelist, Captain Whyte-Melville. He contended that the contractility of the muscles was a property conferred upon them by the nerves. A grand discussion took place between the Hallerians and their opponents; it was carried on for years with keenness and, as scientific men feel strongly on the questions on which they differ, even with some degree of acrimony. As is often the result of such controversies, both disputants were obliged to widen the basis of their opinions, and to examine more closely and carefully the facts on which they founded their theories. Thus the Hallerians recognised that contractile movement occurred in other tissues than in muscle alone, as in the coats of the arteries and in the skin; whilst, on the other hand, their opponents were led to study more carefully the physiology of the nervous system. Thus the dispute contributed to clearer physiological ideas apart from the real question at issue.

Time has declared in favour of Haller. His opponents naturally pointed to the wasting of the muscles after division of the nerves supplying them; but, on the other hand, it has been shown that this is due to the imperfect nutrition which follows a state of inaction, and that if the nerve or muscle be directly and systematically stimulated, the muscle may not undergo very much degeneration nor become less irritable. We know also that we may exhaust a nerve so that when shocks are applied to it the muscle with which it is connected will not respond. If we then send the shock directly to the muscle it may still contract. Again, the protoplasm of plants, and of many of the lower forms of animal life in which no nervous tissue can be detected, manifests contractility.

Let us examine this diagram showing a

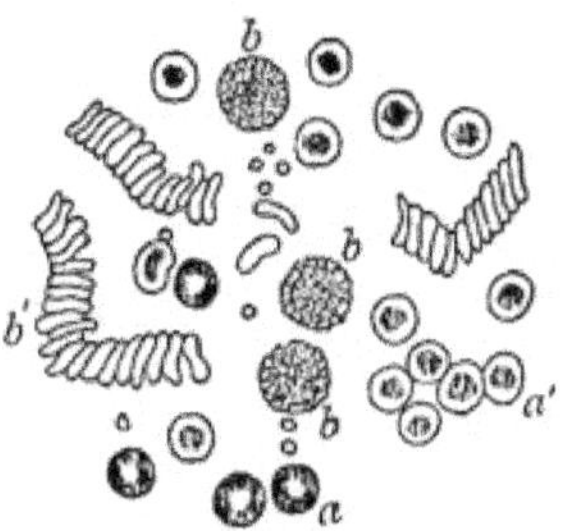

FIG. 41.—Drop of human blood. *a*, coloured or red corpuscle, showing clear spot in centre, or *a'*, dark spot in centre, according to focus; *b*, *b*, *b*, colourless corpuscles, or leucocytes; *b'*, red corpuscles in rouleaux. 300 diameters.

drop of human blood, as seen by a microscope magnifying about three hundred times. You

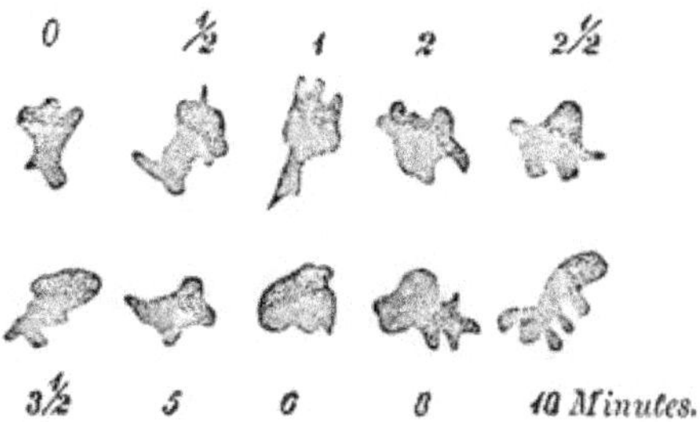

FIG. 42.—White blood corpuscles in frog's blood, magnified 560 times. Changes of form observed during ten minutes. Numbers represent minutes from the beginning of the experiment.

see the red corpuscles in great numbers, but here and there you will observe a few larger ones. These are leucocytes or white blood corpuscles. Careful observations, with high microscopic powers, show us that these are little living things capable of moving and of changing their form, especially when irritated.

Covering many parts of the bodies of some of the lower animals, and abounding on some surfaces, as in the air passages of our own

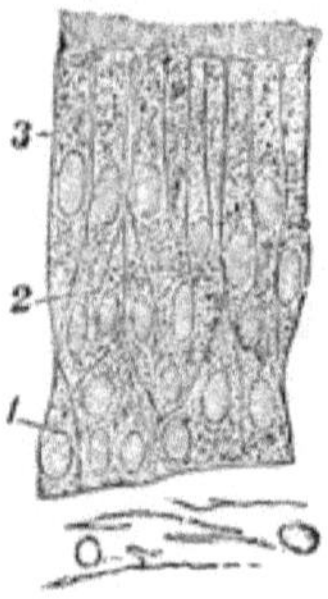

FIG. 43.—Stratified ciliated epithelium. 1, oval; 2, spindle-shaped; 3, cylindrical cells, magnified 560 diameters. From the lining of the nose in the respiratory region.

bodies, are delicate hair-like things called cilia. This diagram shows them as found in the human windpipe. These cilia are destitute of nerves, and yet we see them during life in rapid movement. With high powers, one can see the protoplasm of which they are composed apparently pulsating. All we can say as to their movement is that they show rhythmical contractions. Inherent in their protoplasm is this property of contractility. This statement, you will observe, explains nothing.

Muscular tissue is found in the body in two varieties of fibres which we term the smooth and the transversely striated. Those who examine with their microscopes the structure of the tissues of the body speak of minute things called cells. A cell is a little bit of living

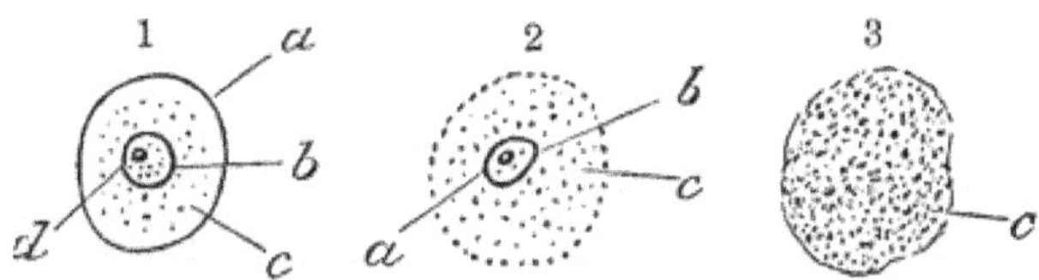

Fig. 44.—1. Original conception of a cell. *a*, cell wall; *b*, nucleus; *c*, cell substance or cell contents; *d*, smaller body, called a nucleolus. 2. Cell wall has disappeared. *b*, nucleus; *a*, nucleolus; *c*, cell substance or contents. 3. Modern view. Cell now consists of granular matter often showing a delicate network of fine fibres.

matter, often having a membrane round it, and almost always having in it a small body, like a kernel, called the nucleus. Cells, always microscopic in size, so that many thousands might be packed away in a space the size of a pea, have many forms, as you see in these diagrams. Some are round, others polygonal, others elongated or drawn out, and others have little processes standing out from them. Now both kinds of muscular fibres are cells the bodies of which are greatly extended lengthways. The smooth fibre-cells are long, spindle-shaped,

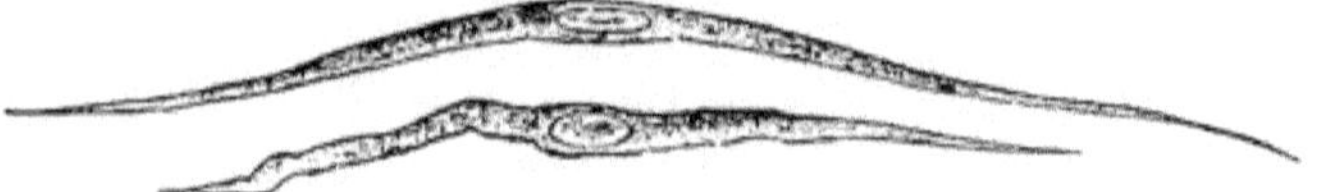

FIG. 45.—Two smooth muscular fibres from the small intestine of a frog.

somewhat cylindrical, bodies with pointed ends, about one-two-hundredth of an inch long by one-five-thousandth of an inch in breadth.

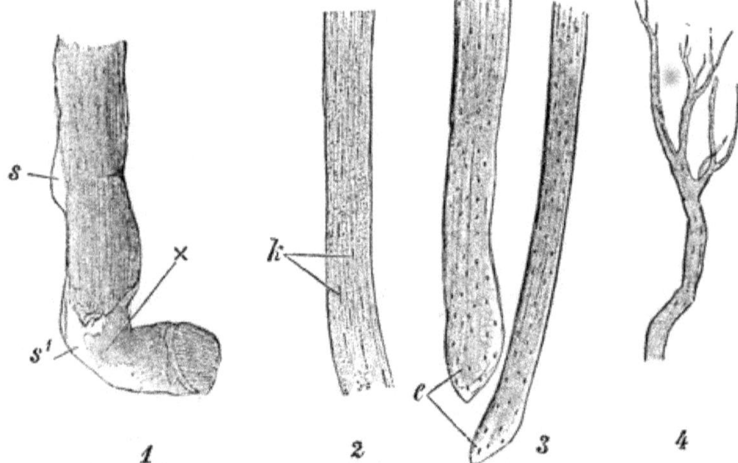

FIG. 46.—Striated muscle of frog. 1, effects of water; *s s'*, sarcolemma. 2, action of acetic acid, showing nuclei at *k*. 3, action of caustic potash. 4, branched muscle-fibre from the tongue. All magnified about 50 diameters.

This kind is found in the coats of the hollow organs, like the stomach, and it is not subject to voluntary control. The other is called striated muscle, because it is apparently formed of fibres, the surfaces of which show markings or lines running transversely across the fibre, as you see in this diagram. Each fibre is really an enormous cell, about one and a half inch long by about one-three-hundredth of an inch broad. The wall of the cell is a fine membrane called the sarcolemma, and in it we find the muscle-substance showing alternately dark and narrower and clear and broader bands. The distance between two of the dark bands is on an average about the one-ten-thousandth of an inch. These dark and light bands are really the edges of discs, so that we have a disc of light substance alternating with a disc of dark substance. Each clear disc, however, when the muscle-fibre is looked at with a high power, shows a thin dark line passing through it, as shown in this diagram, and there is a fainter line, not so well marked, in the centre of the dark disc. Sometimes the fibre splits crossways into discs, and at other times into fine fibrils, each of which shows the same transverse markings as are seen in the fully formed fibre. The dark portions, usually called the sarcous elements of Bowman, after their discoverer Sir William Bowman, so well known in this Institution, are the parts of the fibre that show contractility. By special methods of preparation, Professor Schäfer and others have shown us that the structure of muscle is even more complicated, and that there are peculiar rod-

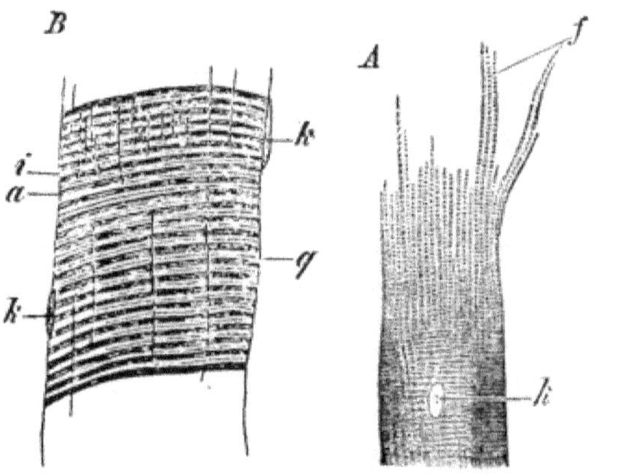

FIG. 47.—B, human muscular fibre magnified 560 times showing the light and dark bands; *k*, nuclei; *g*, Dobie's line. A, the end of the muscular fibre of a frog magnified 240 times splitting into fibrils, *f*; *k*, nucleus.

like bodies having little knobs at their ends running longitudinally in the fibre. These are not seen in the diagram (Fig. 47). It would only weary you to attempt to explain the theories held by histologists (those who endeavour to investigate the nature of tissues) as to the real nature of a muscular fibre. Suffice it to say that it is a very complex structure, containing apparently contractile matter in the form of discs alternating with discs of a substance that is not contractile but which may

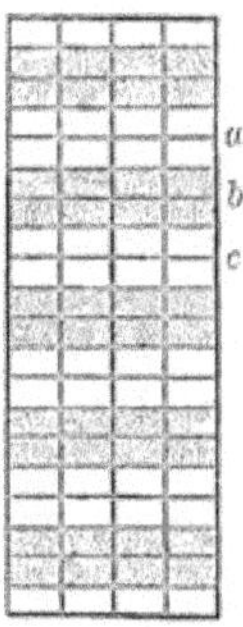

FIG. 48.—Diagram showing hypothetical views as to the structure of striated muscle. Four fibres side by side. *a*, clear bands or discs, each formed of two clear bands or discs, termed the lateral discs of Engelmann, separated by a thin dark line or band, known as Dobie's line, or Krause's membrane; *b*, two discs of dark substance, forming the sarcous elements of Bowman, having in the centre an ill-defined band or line, the median disc of Hensen; *a* is singly and *b* is doubly refractive as regards light.

possibly be elastic. Further, during life, the whole of this remarkable structure is semi-fluid, and there are good grounds for believing that the contractile action is due to the creation of currents passing from one part of the fibre to another, accompanied by chemical changes of a very complicated kind.

As one would expect, the fibres of muscle are intimately connected with a nerve. But what is a nerve? This little white cord, the sciatic nerve of a frog, that we have been experimenting with, consists of numerous delicate fibres called nerve-fibres, each about the one-twelve-hundredth of an inch in breadth.

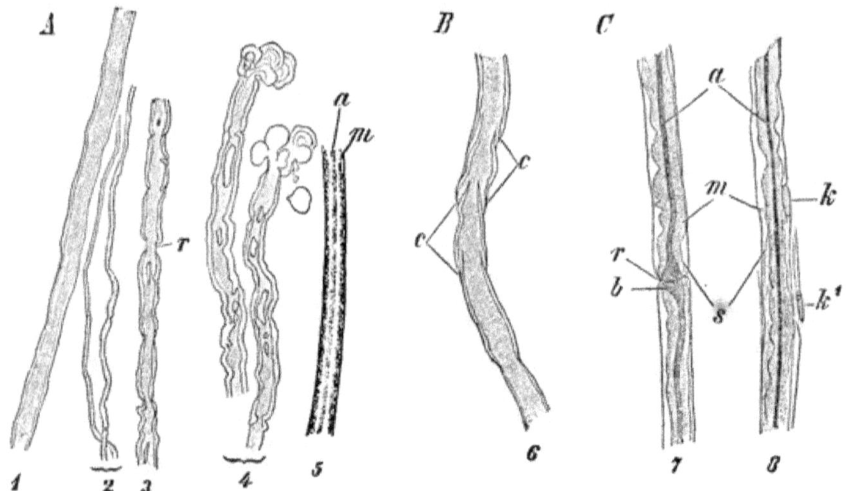

FIG. 49.—Medullated nerve-fibres from the sciatic nerve of a frog. 1, 2, 3, fresh, in a solution of common salt. 3, fibre with a constriction at *r*. 4, fibres as affected by water. 5, as acted on by alcohol. 6, fresh; *c*, segments. 7, 8, hardened; *a*, axis-cylinder; *b*, swelling; *k k'*, nuclei; *m*, white substance; *r*, constriction; *s*, white substance shrinking from neurilemma.

As a rule, each individual fibre has an external sheath called the neurilemma. Inside this we find a cylinder of matter of a fatty nature, known as the white substance of Schwann, and in the cylinder, a core of another substance called the axis-cylinder. These substances during life are semi-fluid. If we traced a nerve-fibre to the brain or spinal cord, we would find it starting from a process, or, as we call it, a pole, of a nerve-cell. When we trace it to a muscular fibre we find it loses the white substance of Schwann, and the axis-cylinder of the nerve-fibre pierces the sarcolemma, or sheath of the muscle-fibre, and ends in what

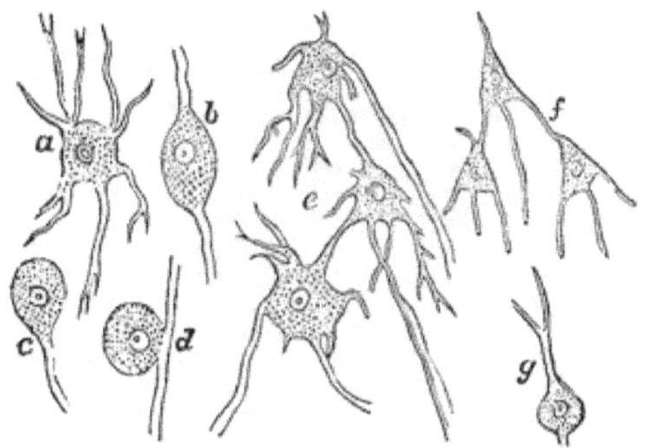

FIG. 50.—Various forms of nerve-cells. *a*, multipolar, from the gray matter of the spinal cord ; *b*, *d*, bipolar, from ganglia on posterior roots of spinal nerves ; *c*, *g*, from cerebellum ; *e*, union of three cells ; *f*, union of cells by processes.

is called an end-plate. The end-plates, seen in the diagrams (Figs. 51 and 52) vary in form and general appearance. Sometimes they consist of very slender fibres, produced by the splitting up of the axis-cylinder, and forming a network ; but usually they take the appearance of irregularly shaped granular masses or discs.

As a rule, each muscle-fibre has a corresponding nerve-fibre. The number of nerve-fibres must therefore be enormous, and a recognition of this fact gives rise to several curious considerations which I shall discuss when we come to consider the electric fishes.

Now we are in a position again to approach the question whether or not the irritability

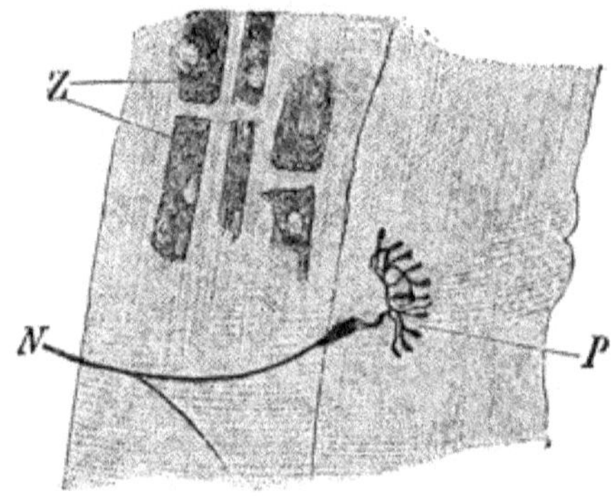

FIG. 51.—Motor nerve-ending from the muscular fibre of the intercostal muscle (muscle between the ribs) of a hedgehog. N, nerve; P, end-plate; Z, flat connective tissue-cells.

of a muscle is inherent in the muscle-fibre. Suppose we could eliminate altogether the nerve-fibres in a muscle, would the muscular substance then contract if we irritated it? From what I have told you, you will admit that we could not mechanically remove all the nerve-twigs from a muscle. They are too small to be manipulated by the most dexterous use of scissors and forceps. We shall have recourse, however, to the action of a substance called curare, which paralyses the end-plates in the muscle-fibres. Here are two muscle telegraphs. The one is connected with a muscle affected by curare, the other with a muscle in a normal state. I have arranged

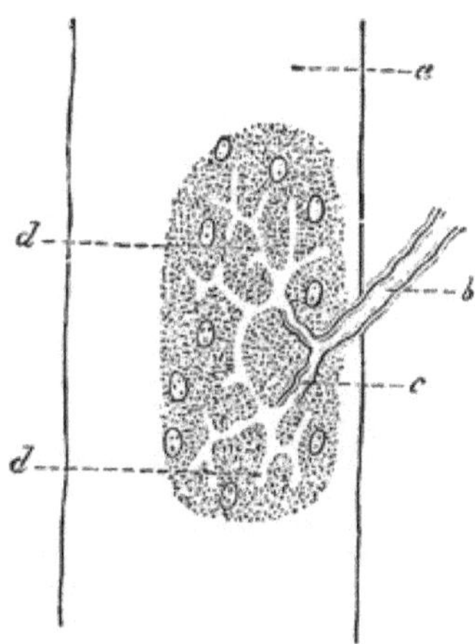

FIG. 52.—A muscle-fibre, *a*, from a lizard. *b*, nerve-fibre terminating in an end-plate.

my apparatus so as to be able to stimulate both nerves at once or both muscles at once. Now I send the shocks from our induction coil to both nerves, and you observe that only one muscle has contracted, moving its telegraph signal; that is, the muscle the nerve of which is not under the influence of the curare.

The other muscle does not respond because its nerve, poisoned by curare, is practically dead, or at all events it cannot act on the muscle. By the curare we have poisoned every nerve filament and every end-plate in this muscle, but we left the muscle-substance just as if we

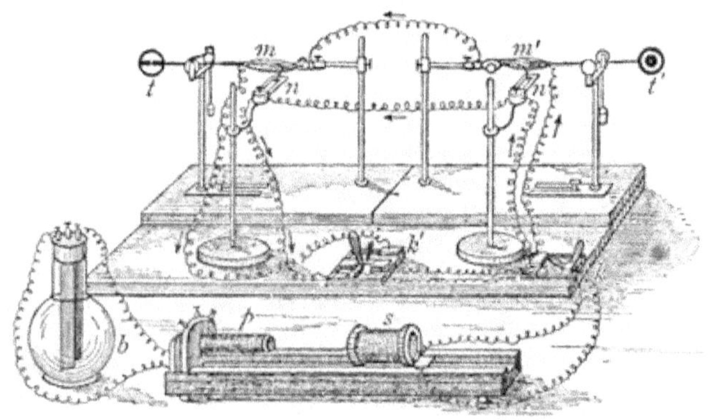

FIG. 53.—Diagram showing the arrangement of the apparatus in demonstrating by curare the inherent irritability of muscle. *b*, galvanic element; *p* primary and *s* secondary coil of induction machine; *k*, key for admitting shocks either to nerves or muscles; *k'*, double key for sending shocks either to the two muscles, *m* and *m'*, or to the two nerves, *n* and *n'*; *t t'* telegraph signals. The arrows show the direction of the currents.

had dissected out and removed all the nervous structures. Now I send the shocks to both muscles, and you notice that both telegraph signals are raised, apprising us that both muscles have contracted. The muscle-substance then contracts when directly stimulated without the influence of nerves. This beautiful experiment was devised by the eminent French physiologist, Claude Bernard, and conclusively proves the truth of Haller's theory.

A muscle, then, contains contractile stuff, and this stuff is thrown into molecular activity by the nerve. The nerve-current, or nerve-shock, is the natural stimulus of the muscle. So long as the muscle is irritable, it responds to this stimulus, and the obvious response it makes is a change of form, or a contraction.

LECTURE IV

Action of a nerve—Rapidity of nerve-current—Nature of nerve-current—Analogy with electric current—Nerveless animals—Heat in muscle—Muscles liberate mechanical energy and heat—Chemical changes in muscle.

We have now seen that the living matter forming a muscle is irritable or excitable, that is to say, it responds or reacts to a stimulus. We have also learned that the muscle shows its response or reaction by a contraction. Lastly, we have found that the natural stimulus that sets the muscle into action is something that happens in a nerve. Let us to-day, in the first place, study more carefully than we have yet done what occurs in a nerve.

A nerve, like a muscle, is composed of living matter, and this living matter, like all living matter, is irritable; but it does not show its irritability in any way evident to our senses. Suppose I irritate a little bit of nerve, which I have every reason to think is still alive, it shows nothing. The electric shocks sent to it produce no evident result. We must not assume, however, because we see nothing following the irritation of a nerve, that nothing happens. There may be changes in the nerve for all we know to the contrary, and the fact that changes do occur in the nerve would be evident if the nerve had still been connected with a muscle, because then, as you now know, irritation of the nerve would have been followed by contraction of the muscle. We would have seen the muscle move, and that would have been a proof that something really occurred in the nerve at the point of irritation, and that something passed along the nerve to the muscle. But perhaps you think I am going a little too fast. You may say that it is possible that irritation of the nerve at one point causes an instantaneous change throughout all the nerve, and that nothing really *passes* along it. Now this is a question that we can only settle by experiment.

Suppose that we irritate a nerve close to where it enters a muscle, the muscle will not contract at the instant the stimulus is applied to the nerve. There is always a loss of time.

We may suppose that this time may be divided into three portions. First, a period in which changes occur in the nerve; second, a period in which changes occur in the muscle (the latent period); and third, a period occupied by the contraction of the muscle. Now suppose that instead of irritating the nerve close to the muscle, we irritate it at a point farther off, say two inches from the muscle. If, then, the times of the latent period and of the contraction remain the same, and if something travels along the nerve from the distant irritated point, the muscle should contract a little later than when the nerve is irritated close to the muscle. Reasoning in this way, Professors von Helmholtz and du Bois Reymond, now a good many years ago, devised methods by which this experiment may be made.

Let us try an experiment or two to illustrate this method. One of the most ingenious and simplest instruments for the purpose is the spring myograph of Professor du Bois Reymond, which I now show you. It consists of a smoked-glass plate, which is driven in front of the recording marker of the myograph by the recoil of a steel spring C. Underneath the frame carrying the glass plate are two binding screws at F, to one of which is attached an arm of brass 1, which can so move horizon-

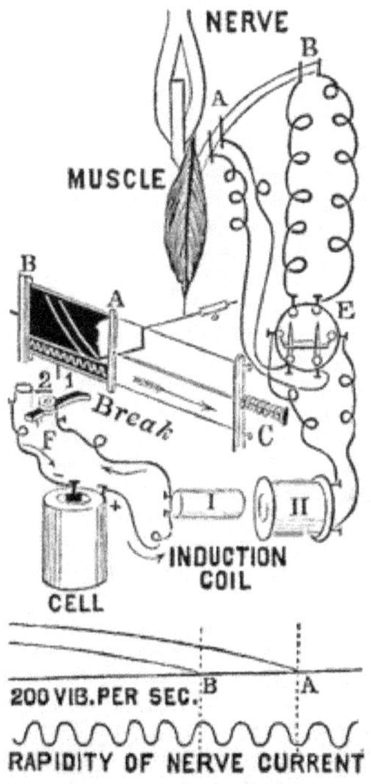

FIG. 54.—Diagram showing arrangement of apparatus in the experiment of measuring the rapidity of the nerve-current. For description see text.

tally as to establish metallic contact between the two binding screws marked 1, 2. By means of these screws the myograph is interposed in the circuit of a galvanic element and the primary coil I of an induction machine, and the brass arm is so placed as to connect both binding screws, thus completing the circuit. From underneath the frame carrying the smoked-glass plate there descends a small

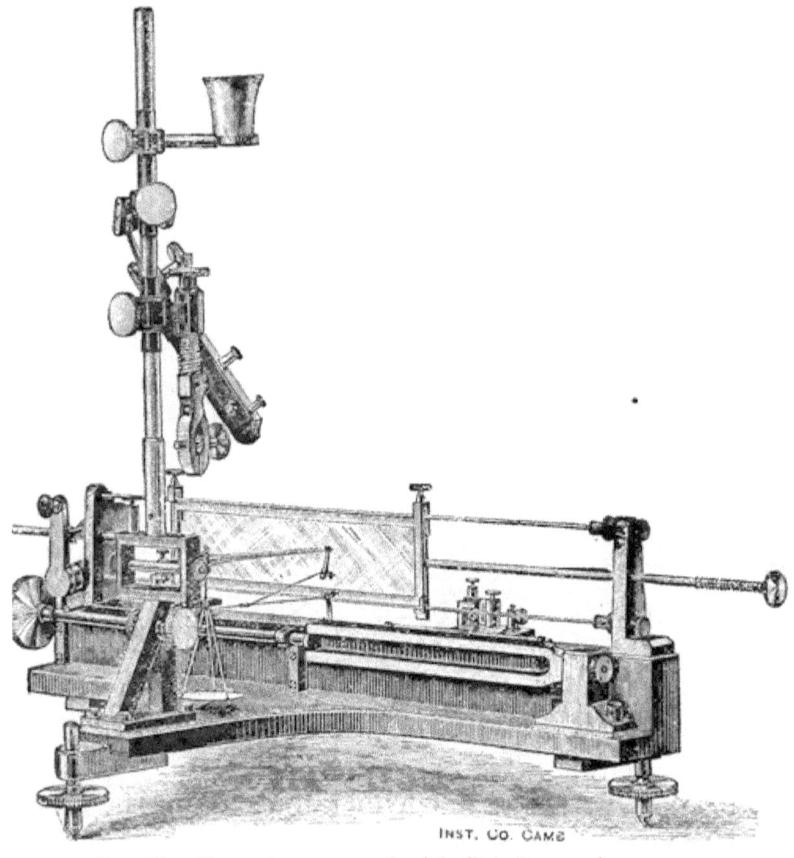

FIG. 55.—The spring myograph of du Bois Reymond.

flange, which (when the glass plate, by releasing a catch not seen in the figure but close to C, is driven across by the spiral spring from left to right) pushes the brass arm aside, and thus breaks the circuit of the primary coil. When this occurs an opening shock is sent from the secondary coil II to a commutator, E, an instrument by which electric currents may be transmitted to the nerve, either to a point close to the muscle at A, or at a distance from it, B.

Now we have the apparatus arranged so as to send the shock to the nerve at a point close to the muscle A; the muscle contracts, and draws by means of the marker, on the smoked surface of the glass, the curve seen at A in the lower part of the diagram. This leaves the horizontal line (which would be drawn by the marker were the muscle at rest) at A. We shall, in the next place, arrange for another experiment, in which the nerve will be stimulated at a distance from the muscle, at the point B, in the upper part of the diagram. This we do by again pushing the smoked-glass plate back to its first position, closing the primary circuit by the brass arm at the binding screws, and reversing the commutator so as to send the shock along the wires to B. Touch the spring; the plate again darts across, breaks the circuit, and the muscle again contracts, but this time it describes on the smoked surface the curve B, seen to the left of A, in the diagram. You observe this curve leaves the horizontal line at B, that is, a little *later* than when the nerve was stimulated close to the muscle (Fig. 54).

It follows, therefore, that the distance on the horizontal line from A to B represents the time occupied by the transmission of the nervous impulse from B to A of the nerve. We measure the rate at which the glass plate was travelling by bringing to bear on it a marker connected with one of the prongs of this tuning-fork, and we cause the fork to vibrate at the moment when the glass plate dashes past the markers (Fig. 55). The time waves thus accurately measure the rate of movement of the glass plate, and consequently the minute interval of time between A and B.

This experiment proves that when a nerve is irritated at any point some kind of change is then produced, and that a change is propagated with a certain velocity along the nerve. This something we call a current, for want of a better term ; but it is not a current but something sent on from point to point. It travels slowly compared with the velocity of light or of electricity. In the nerves of the frog the velocity is about eighty-seven feet per second, and in higher animals of constant temperature, such as in man, it only reaches a speed not exceeding three hundred feet per second.

The real nature of the change in a nerve-fibre during the transmission of the "current" is unknown. A nerve is both a receiver and a conductor of impressions. It can be stimulated at any part of its course, and from the stimulated point something is propagated along the nerve. Many explanations have been offered, but none is satisfactory. Naturally one thinks of the passage of electricity along a conductor, but, as I have said, the current is incomparably slower than the passage of electricity even along a nerve. The appearance of a nerve-fibre with its axis-cylinder is not unlike a wire insulated by some substance like silk or gutta-percha. Wires which direct the electrical change are insulated for the purpose of preventing the electricity from passing from one wire to another. We have no evidence that the nervous change can pass from one nerve-fibre to another. We know also that when an electric current passes along one wire it may produce currents, so-called induction currents, in adjacent wires ; but there is nothing analogous to this in nerves. We do not know of induced nerve-currents. Each nerve-fibre appears to conduct its own change or current.

The phenomenon is more like that of a rapid series of chemical changes passing quickly along a tract, as when a train of gunpowder slowly burns, or when a long thin band of gun cotton, such as we have here, is seen to burn slowly from end to end. But the analogy is not complete. The train of gunpowder and the band of gun cotton disappear and leave nothing behind, but the nerve-fibre remains. It must be said that the evidence we possess of chemical changes in the nerve-fibre is very meagre, no doubt because of their comparative insignificance. Still, small as the change is, it is sufficient to set off the highly unstable material in a muscular fibre and to produce chemical changes attended by the liberation, as we have seen, of mechanical energy.

The change in a nerve-fibre can also produce changes in other organs. If the nerve-fibre reaches the cells in the spinal cord it may set up changes in these which result in a transmission of nerve-currents or shocks along other fibres, as in the phenomena of reflex action. Again, if the fibre passes to the brain, it excites changes in nerve-cells connected in some way with consciousness, and we thus come to know of something which affected the fibre at its commencement. Thus when light falls on the eye it affects the nervous structure called the retina; the retina is connected with the brain by nerve-fibres which are affected by the changes occurring in the retina, and nerve-currents travel along these nerve-fibres to the brain. In the brain they set up changes in nerve-cells which result in the consciousness of light, that is to say, we have a sensation which we call light. In all these instances, the nature of the change in the nerve-fibre and the mode of its transmission are the same. The results are different because the fibres end in different kinds of terminal structures.

Thus an electric current travelling along a wire may do very different things according to the nature of the apparatus at the end of the wire. Here is a wire conducting a powerful current. At this point, we cause it to branch out so as to divide the current into a number of streams. You see here the current decomposing water, there magnetising soft iron, here again doing the mechanical work of turning a wheel. In like manner, there is contraction of a muscle if the nerve ends in a muscle, change in the calibre of a blood-vessel if the nerve ends in that structure, secretion from a gland if the nerve is connected either with the vessels or the cells of a gland, an electric discharge or shock if the nerve terminates in the electric organ of an electric fish, and a feeling or sensation if the nerve-fibre goes to a sentient brain.

But if nerves are of so much importance, you will naturally ask how motions are produced in animals that have no nerves. Many animals show not a trace of nervous structures and yet they move. Again, the hearts of some animals beat with great regularity and still no nerve-fibres exist in their tissue. Such nerveless structures respond to a stimulus. Give them a shock of electricity and they contract. There can be no doubt that in such cases the contraction of one part produces some kind of disturbance, it may be electrical, which is propagated to adjoining parts, and acts upon these as a stimulus. Thus a kind of wave of contraction passes through the structure. Something of the same kind has been observed in muscle; but to this we shall return when we come to the consideration of the electrical phenomena in muscle.

We have now studied muscle as a producer of what we may call mechanical energy. At this stage, I shall leave the order in the syllabus and take up a subject mentioned in connection with the fifth lecture, namely, the production of heat by a muscle. We detect heat usually by a thermometer; but the heat we must look for in a muscle is so small in quantity as to oblige us to use a more delicate method. It is well known that minute quantities of heat may be detected by the use of what are called thermal piles. To understand the principle involved in the working of a thermal pile, look at this simple experiment. I have here a number of strips of the metals iron and copper soldered together. The points at which the two metals are fixed together are called junctions. The apparatus is put into connection with a galvanometer, and you will observe the coils of the galvanometer contain only comparatively few coils of wire. Such a galvanometer is said to be of low or small resistance, and it is well adapted for

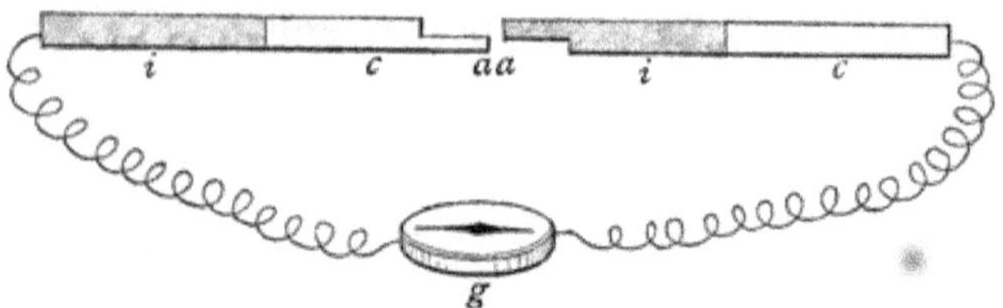

FIG. 56.—Thermal junctions of iron, *i*, and copper, *c*. *g*, galvanometer; *a a*, junctions that may be pressed together.

such an experiment as we are now about to make. Now if I heat one set of the iron and copper junctions by simply pressing them together at *a a* (Fig. 56) while the other set is kept cool, a current of electricity is generated which passes round the coils of the galvanometer and causes a deflection of the needle. A very small difference in the temperatures of the two sets of junctions is quite sufficient to produce a current.

This arrangement is made more sensitive by having a large number of thermo-electric junctions constituting a thermo-electric pile. I shall connect this pile with our galvanometer, and you see it is so sensitive that, if I hold my hand near it, the heat radiating from my hand at once causes a movement of the galvanometer needle and a corresponding movement of the spot of light on the scale. When I heat the other set of junctions, you observe the movement of

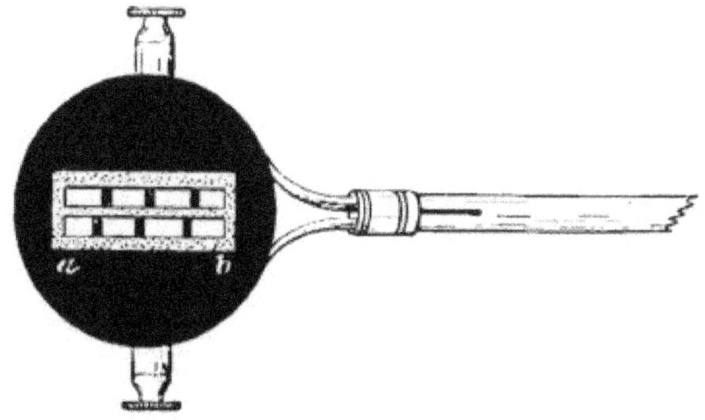

Fig. 57—Small thermal pile. Observe the junctions of bismuth and antimony. The distance from *a* to *b* is one-fourth of an inch.

the spot of light is in the opposite direction, because the current passes in the opposite direction through the coils of the galvanometer.

Let us now examine the muscle by a thermo-electric arrangement. I have two very small thermal piles, and I connect them together, so that if I heat one by bringing my hand near it, the spot of light from the galvanometer moves to the right, and if I heat the other it moves to the left. We now place this one, which causes a deflection to the right when it becomes hot, in connection with a muscle to which a weight is attached, and we place the nerve of the muscle over the two wires from the secondary coil of our induc-

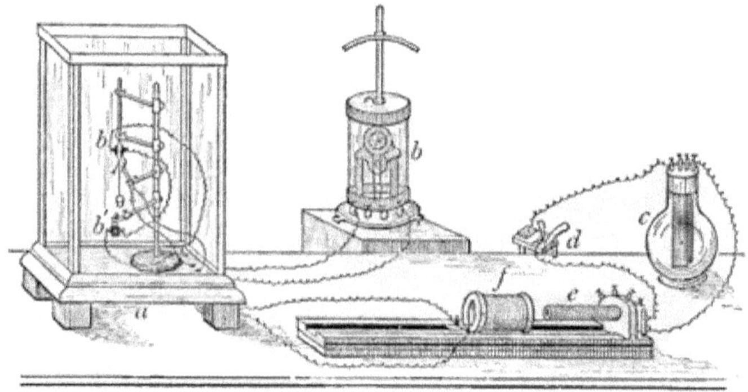

FIG. 58.—Diagram showing the arrangement of the apparatus in the demonstration of the heat of muscular contraction. *a*, case with sides of thick plate glass; *b b'*, thermal piles; *b*, galvanometer; *c*, galvanic element; *d*, key; *e* primary and *f* secondary coil. The contents of the glass case are seen on a large scale in next figure. The galvanometer has a low resistance.

tion machine. To keep off all radiant heat as much as possible, we shall enclose the whole apparatus in a square chamber, the walls and roof of which are made of plate glass one inch in thickness. Notice the position of the spot of light. I now open the key so as to tetanise the muscle, and you notice that at once the spot of light moves to the right, proving that the thermal pile with which the muscle was in contact has become hotter, or, in other words, that the muscle in the tetanic state has be-

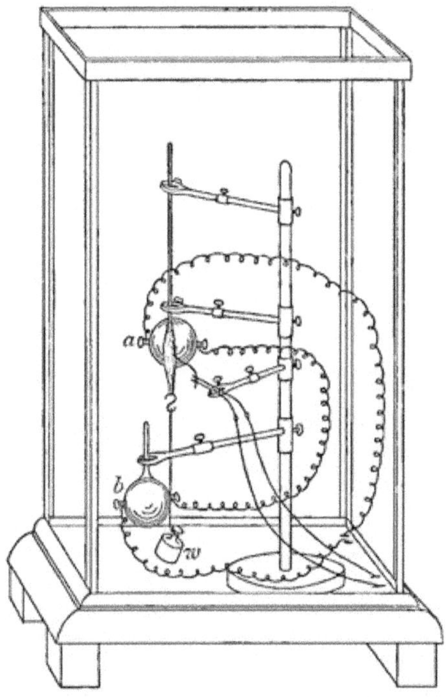

FIG. 59.—*a*, thermal pile touching muscle ; *b*, other thermal pile ; *w*, weight keeping muscle on the stretch.

come hotter. It is not so easy to show that even a simple contraction produces heat, because such a movement of the gastrocnemius of a frog is associated only with a rise of temperature of from one-thousandth to one-five-thousandth of a degree centigrade. Tetanus of frog's muscle gives from fourteen to eighteen-thousandths of a degree centigrade. No doubt each contraction of the muscles of higher animals, such as those of man, produces or is associated with more heat, but still the amount for each individual contraction is not much. But by the accumulation of small amounts, a large amount is formed, and there can be no doubt that a large proportion of the heat of our bodies is derived from the muscles.

That muscles produce heat is consistent with our daily experience. The more actively we work our muscles the hotter we become. When we wish to become warm we run, or leap, or dance, and in doing so we exercise our muscles.

We must now take a more scientific view of this matter. Muscles liberate energy as mechanical energy and as heat. The mechanical energy does work by moving one part of the skeleton upon the other, or by lifting a weight, as when I lift this book. Heat is another mode of energy, and we have seen that it is also liberated. Now a steam-engine does the same kind of thing. It expends or produces mechanical energy, and it also becomes hot. But we know that energy never comes out of nothing. You cannot get it for nothing. You can only get it by the expenditure, or, if you put it in another form, by the disappearance, of another mode of energy. This is one of the greatest thoughts of modern times—this thought of the persistence of energy. We cannot create or destroy matter. We can only transform it. In like manner, we cannot create or destroy energy. We can only transform it. Our steam-engines, and gas-engines, and hot-air engines are all transformers of energy. None of them makes it; they receive it from something, and they pass it on in other forms.

Take a steam-engine. It works by the steam expanding by heat and moving the piston. The hot steam comes from the boiler containing the water. The water is heated by the fire of the furnace. In the fire combustion or burning is taking place. The fuel, consisting chiefly of matters rich in carbon, is burnt, that is to say, the carbon unites with the oxygen of the air to form carbonic acid gas; but in this chemical operation which we call burning or oxidation, heat is set free, and the heat is the energy that drives the piston. The piston moves and drives all kinds of machinery, that is to say, the heat that moves the piston, through the medium of the steam, is transformed partly into what the engineer would call mechanical energy. It is this mechanical energy that does work. All the energy, however, set free by oxidation from the fuel is not transformed into mechanical energy. A part, a large part, as much as eighty-eight per cent of it, is set free simply as heat, which, I need hardly say, is of no use to the engineer. The same kind of reasoning guides engineers in the construction of all kinds of engines, and they are always striving to get as much as possible of the energy of the fuel transformed into mechanical energy.

Now turn to our muscle. Is it also a transformer of energy? If it is capable of manifesting mechanical energy, as undoubtedly it is in doing the work of lifting a weight, and if it becomes hot, these two energies, mechanical and thermal, must come from somewhere. Living though it be, it can no more create energy than the metallic, dead steam-engine can do.

Can we show that the muscle is also the seat of chemical changes? If we can do so, we may find that there are operations going on in a muscle that are comparable to the combustions or burnings in the furnace of a steam-engine. Let us try. Here is a bit of fresh muscle. I test it with red litmus paper, a well-known test for alkalies, and you observe the paper becomes slightly blue. We find then that the reaction, as it is termed, of a quiet muscle is alkaline. The muscle is alkaline on account of certain alkaline salts of soda present in it. Let us now test a similar muscle that has been tetanised since the beginning of the lecture. You see the red litmus shows nothing. There is no blueness, as in the other case, and we conclude that the muscle is not now alkaline. But we now test it with a bit of blue litmus paper, which is the method employed by chemists to detect acids, and you see it is reddened. This reddening shows the presence of an acid substance, and careful chemical research has proved that the acidity is due to a kind of lactic acid, an acid closely allied to the acid that we find in sour milk, hence called lactic (from *lac*, milk) acid. Here is evidence then of one chemical change produced by or connected with activity of muscle.

We are all familiar with the fact that living things breathe, and that breathing is the taking in of oxygen and the giving out of carbonic acid gas. When we inhale air in inspiration, the air, which is a mixture of two gases, oxygen and nitrogen, passes into our lungs through passages or tubes that become narrower and narrower, until they end in little sacs or dilatations known as the air-cells of the lungs. On the walls of the air-cells are networks of minute blood-vessels in which the blood flows, and it is here that respiratory exchanges occur between the air and the gases that exist in the blood. Oxygen gas passes into the blood and carbonic acid gas passes out. It is not necessary to demonstrate to you that oxygen gas is necessary for breathing. We all know that this gas must be present in any atmosphere fit for breathing, and that if an animal is placed in an atmosphere containing no oxygen, or if it is placed in a vacuum, it very quickly dies. The best test for carbonic acid gas is lime-water, which becomes turbid when the gas is led through it or shaken up with it. The chemists have given me this jar of carbonic acid, and you see

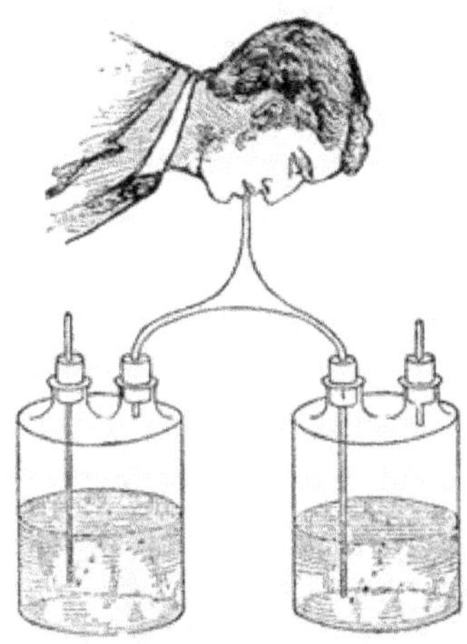

FIG. 60.—Breathing into lime-water. Faraday's method as shown in the *Chemistry of a Candle.*

when I shake up some of this clear lime-water with it, how white and milky it becomes. We can readily, by this test, show you that carbonic acid is produced by breathing, and it is interesting at a Christmas lecture at this Institution to employ the simple method used thirty years ago by Faraday in his celebrated course on the chemistry of a candle. I have two bottles, one containing lime-water and the other common water, and, by the arrangement of the tubes, when I inspire I draw air through the water, and when I expire I blow air through the lime-water. You see the water remains clear but the lime-water becomes turbid in a few minutes. It becomes turbid by the carbonic acid in the breath combining with the lime in the lime-water so as to form carbonate of lime, which is not readily dissolved, and consequently gives the white appearance to the lime-water. Carbonic acid gas, then, comes from our lungs.

The air in the lungs, as I have said, receives the carbonic acid from the blood. Those who are unfamiliar with physiology can hardly conceive of the blood as containing a large amount of gas. Take a hundred cubic centimetres of blood: this quantity may contain about sixty cubic centimetres of gas, and perhaps two-thirds of this consists of carbonic acid, the other one-third being oxygen. Although it contains this large amount of gas, blood does not effervesce in the air, because the pressure of the air on its surface prevents it from escaping at ordinary temperatures, but if we allow blood to run into a vacuum, as I do when I allow it to pass into this large jar, you see how it effervesces.

The question that next arises is, Where does this carbonic acid come from? The blood, as you know, is sent out through the arteries by the force of the heart-beat; these arteries become smaller and smaller until they end in networks of fine vessels called capillaries, many of which are not wider than the one-two-thousandth of an inch; and from these capillaries the veins originate that carry the blood back to the heart. Capillaries exist in greater or less number in almost all the tissues, and it is by the blood circulating in these that the tissues are nourished. Under the pressure in these minute vessels, fluid matter oozes through their walls and bathes all the neighbouring tissues. This fluid holds in solution the matters needed for nourishing the tissues, and it also contains gases in solution. Thus the fluid is both a nutritive and respiratory medium; by it the tissues are nourished, and by it they breathe. Each little element of tissue needs oxygen, and it produces carbonic acid gas. The blood in passing through the tissues thus loses, to some extent, the one gas, oxygen, while it gains the other, carbonic acid. It is thus changed from arterial or bright scarlet blood to dark venous blood. Here are two jars, one containing oxygen the other containing carbonic acid. I add a little blood to each and shake it up with the gas. See the magnificent scarlet of the one and the dark purple of the other. The blood thus made venous is carried back to the heart by the veins, and is then sent to the lungs. Here it gets rid of a good deal of its carbonic acid, and gains more oxygen, and it is thus reconverted into arterial blood, to be again distributed through the body.

Now the tissues in which the consumption of oxygen and production of carbonic acid go on with greatest rapidity are the muscular tissues, and the more mechanical energy a muscle expends in doing work, the more oxygen it needs and the more carbonic acid it produces. The venous blood flowing out of a muscle is always rich in carbonic acid. Here are two muscles, both in an atmosphere of oxygen. Each is in a little tube inverted over mercury. The one has been at rest, the other has been tetanised at intervals by an arrangement I need not at present describe. You observe that the mercury has risen in each case. I add a little lime-water to the tube containing the muscle that has been working hard, and you see how muddy it

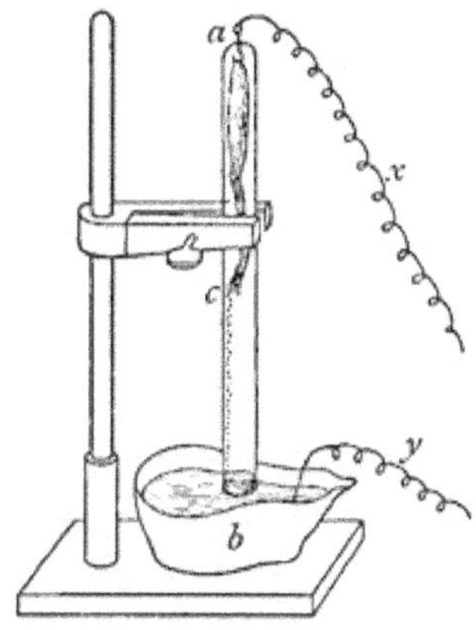

FIG. 61.—Muscle in tube of oxygen over mercury. *a*, platinum wire fused into end of tube and connected with small hook, from which a frog's limb is suspended; *c*, toe of limb; *b*, trough containing mercury. A small amount of mercury is on the side of the tube between mercury in *b* and toe at *c*, so that induction shocks sent in by *x* and *y* readily tetanise the limb. The limb receives tetanising shocks at intervals of thirty seconds, and the experiment may go on for sixty or eighty minutes.

at once becomes. The same experiment with the resting muscle does not show the same degree of muddiness, indicating that the resting muscle has not produced so much carbonic acid as the working muscle.

Still it is interesting to observe that even

the resting muscle breathes. It is a little living thing, taking in oxygen and giving out carbonic acid. When it is obliged to work hard it breathes faster, that is to say, it produces more carbonic acid, and uses up more oxygen. This is exactly what is consistent with experience. Active muscular exercise, as in running, causes an increased consumption of oxygen, and an increased elimination from the blood of carbonic acid.

I imagine some of you may be asking the question, Will a muscle contract only in an atmosphere containing oxygen? Let us appeal to experiment. In this jar, is a muscle still capable of contracting, and yet there is practically no air here, as it has been nearly all removed by an air-pump. In this other jar you see a muscle contracting in an atmosphere of nitrogen, and in this third jar a similar muscle contracting in an atmosphere of hydrogen. You observe this fourth one contracting even in a jar of carbonic acid. It is quite true a muscle will not live long in these circumstances, but even in an atmosphere containing no oxygen a muscle will go on producing carbonic acid. Now as carbonic acid is a compound made up of carbon and oxygen, it is evident that the muscle must have got oxygen from somewhere. The only thing we can say about this is that the muscle takes it from matters in its own substance that contain oxygen.

We have learned, then, that an active muscle becomes acid, that it uses up oxygen, and that it produces carbonic acid. Other chemical changes happen in a muscle that I will not attempt to demonstrate, as the methods by which this can be done require time and many refined appliances not suitable for the lecture-room. I shall merely mention them. Thus a peculiar kind of starch (glycogen), formed in the liver, is carried to the muscles by the blood, and is there consumed. We do not know, however, how the muscle uses the glycogen, whether it uses it directly or whether it first splits it up and then uses some product of its decomposition. So long ago as 1845, von Helmholtz pointed out that by exercise the substances that can be dissolved out of muscle by water are diminished, while those soluble in alcohol are increased, indicating that the one set of substances was used up, while the other set was probably produced by muscular activity. It is also well known that very complex bodies containing nitrogen are found in extracts of muscle, and it is highly probable that these are in a way waste products—substances that have resulted from the breaking down of the matter of the muscle. These facts that I have laid before you all point one way. They all tell of chemical changes in muscle, and they all support the statement that the harder the muscle has to work, the greater is the activity of the chemical phenomena happening in it. Let us put this in more correct scientific language. The setting free of energy by the muscle,—as mechanical energy when it moves, and as heat when it becomes warm,—is associated with, and is likely the result of, the chemical processes happening in it.

LECTURE V

Source of muscular energy—Food stuffs—Milk—Muscle as an engine—Repair of muscle—Ratio of mechanical energy and heat—Fatigue of muscle—Athletics.

Before beginning this lecture, I will show you an experiment which Professor du Bois Reymond calls the muscle dance. You see the muscle connected with the interrupter (see Fig. 11, p. 30), so that when it contracts it breaks the primary circuit of the induction coil and also rings the electric bell. On the scale-pan below the interrupter, I have placed a heavy weight, which at once stretches the muscle when it begins to relax. The relaxation of the muscle, however, again closes the primary circuit, and the muscle receives a shock from the secondary coil. This again causes it to contract and to ring the bell. Again it relaxes and gets another shock. Thus the muscle breaks the circuit by its contraction, and forms it by its relaxation, and each time it does so it gets a shock. It thus dances to its own music, as you now see and hear.

We have hitherto studied muscular movements caused either by a direct electrical stimulus or by the action of a nerve. There are, however, rhythmic movements of muscular substance. You see here a frog's heart attached to an apparatus by which it can be fed with blood, and you see it beating with great regularity. The heart is caused to work a little manometer, and on the mercury in one limb of the manometer I have placed a little glass rod bearing a flag, so that you may see the flag moving up and down with each beat of the heart. In this way a heart can be kept alive for hours, and we can estimate the amount of work it is able to perform. This also illustrates a method by which physiologists can examine the influence of substances on the heart. Thus we might feed the heart with blood free from any poison and note how it worked. Then we might feed it from this other tube with blood containing a small percentage of the substance to be examined, and again note the effect. By comparative

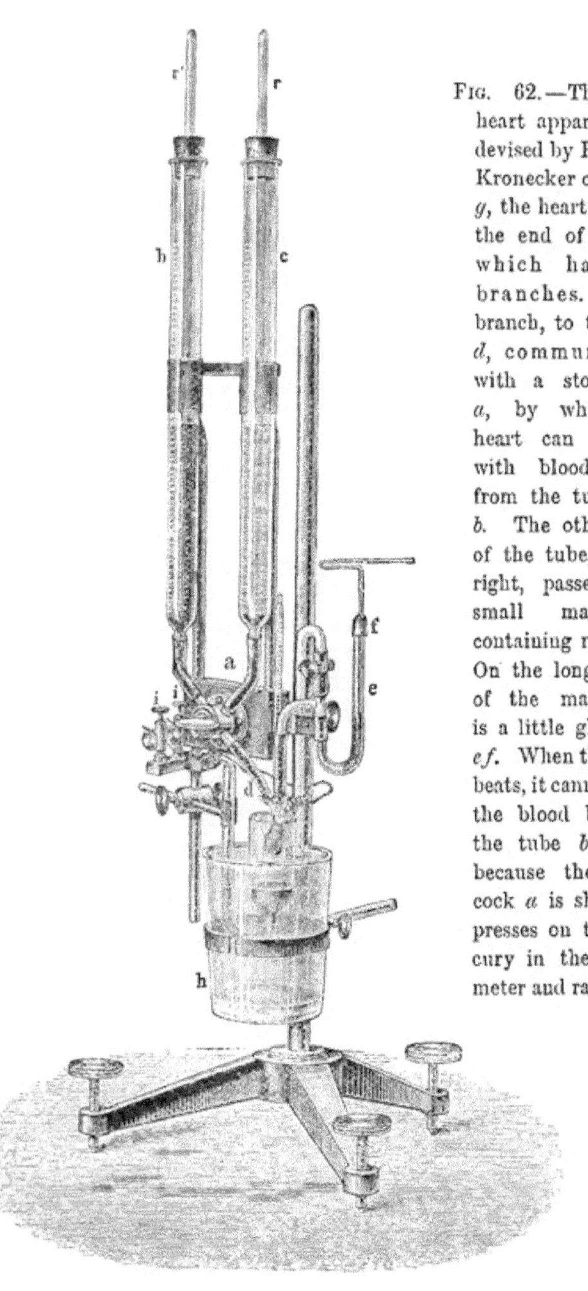

FIG. 62.—The frog-heart apparatus, as devised by Professor Kronecker of Berne. *g*, the heart fixed on the end of a tube which has two branches. One branch, to the left, *d*, communicates with a stop-cock, *a*, by which the heart can be fed with blood either from the tube *c* or *b*. The other limb of the tube, to the right, passes to a small manometer containing mercury. On the longer limb of the manometer is a little glass rod, *e f*. When the heart beats, it cannot force the blood back to the tube *b* by *d*, because the stop-cock *a* is shut. It presses on the mercury in the manometer and raises *e f*.

experiments of this kind, we can get valuable information of great use to the physician, and it is satisfactory to know that information thus obtained from experiments on the frog's heart has been found to agree with that got by experimenting on the hearts of warm-blooded animals. This is another example of how valuable to humanity the frog has been in the way of giving scientific information. It is not too much to say that each time your physician sees you when you are ill he brings to the study of your case knowledge that has been gathered for him by the physiologist from the frog. As we are in the habit of commemorating by monuments the services of those who have been benefactors to humanity, I know no animal,—no tiger, lion, or panther,—that better deserves a bronze statue than the humble frog. Such a statue in Trafalgar Square or on the Thames Embankment would not inappropriately mark our appreciation of the services he has involuntarily rendered to humanity.

Referring to rhythm, look for a moment at this experiment in which you see a muscle—the sartorius—beating rhythmically like a heart, because it is immersed in a fluid, called Biedermann's fluid, composed of common salt, alkaline phosphate of soda, and carbonate of soda and water. This striking experiment, which always is of great interest to me, shows how rhythmic movement may to some degree depend on nutritional changes going on in the muscle.[1]

We saw in last lecture that a muscle is a little laboratory in which chemical processes go on, and that the energy manifested by the muscle depends upon the activity of these operations. If a muscle is constantly throwing off effete matters arising from the wear and tear of its substance, and if it is always expending energy, fresh matter and fresh energy must be supplied to it. What is the source of supply? You naturally answer, the blood, and this answer is right. It is this fluid that brings to the muscle the matters that it uses to build up its substance or the matters that it acts upon, as we may suppose a machine acting on something supplied

[1] Biedermann's fluid: chloride of sodium, five grammes; alkaline phosphate of soda, two grammes; carbonate of soda, five grammes; water, one litre.

to it. But a further question arises. Whence does the blood receive these new materials? Evidently from the food and from the oxygen we take into the blood by respiration. Food stuffs are then the source of the energy set free by the muscle.

A very little consideration shows us that animals live on food stuffs that are apparently very different from each other. An ox eats grass; a lion lives upon flesh; a man prefers a mixed diet, such as meat and potatoes. The diets of even the various races of mankind present remarkable differences. The native of Bengal lives largely on rice with a little fat; some Europeans, like the French and Italian peasantry, partake of a diet consisting almost wholly of vegetable substances; the Turkoman, in the steppes of Central Asia, consumes vast quantities of flesh; and the Esquimaux finds that a diet rich in fat best enables him to withstand the rigours of his severe and inhospitable clime. Men have found out by experience what suits them, and no doubt custom or habit has a great deal to do with the selection they make. Can we then compare their dietaries with the view of solving the problem how energy is to be obtained from substances so unlike as those I have mentioned? We get light on this question by examining a natural food, one upon which almost all men live in the earlier part of their existence, the food of childhood, milk.

I pour a glass of milk into a basin and warm it, adding a few drops of acetic acid. You see it quickly undergoes a change. Masses of curd make their appearance and float in a fluid of a yellowish colour, familiarly known as the whey. We filter it. The curd you see is a soft friable matter. It consists of a body called casein, of very complex chemical composition. If we gave it to a chemist to analyse, he would tell us that it contained carbon, hydrogen, oxygen, and nitrogen. Note particularly that it contains nitrogen; it is, as we say, a nitrogenous substance, and it represents the first constituent of every diet, which must contain a nitrogenous or, as it is termed, a proteid substance. Proteid bodies include such substances as we find, along with other matters, in white of egg, in meat, in wheat, in oats, in beans, in grass, and in many other well-known articles of diet. The casein, as I have said, represents the first essential constituent of a diet, proteid or nitrogenous matter.

Now let us examine the whey. It is sweet from containing a kind of sugar called milk-sugar. These sugar substances, when heated with an alkaline solution of a salt of copper, have the property of taking oxygen from the copper compound, changing it to one that is insoluble; and in effecting this change they also cause a change of colour. Thus you see, when I add to this solution of grape-sugar a blue solution of a copper salt (called Fehling's solution, an alkaline tartrate of copper) and heat it, the blue colour gradually disappears and a reddish substance, an oxide of copper (cuprous oxide) falls to the bottom of the glass. Applying the same test to the whey, we find proof of the existence of sugar in it. Now sugar is a compound of carbon, hydrogen, and oxygen. It contains no nitrogen, and hence it is called non-nitrogenous, to distinguish it from the nitrogenous or proteid group of bodies already referred to. As the oxygen and hydrogen in sugar are in the proportion by volume of one of oxygen to two of hydrogen, as in water, sugar is said to be a carbo-hydrate; or, as one might phrase it, it is hydrated carbon. Suppose, as I now do, we pour some strong sulphuric acid, which has a great affinity for water, upon a bit of lump sugar, you see the lump soon becomes a black mass of carbon or charcoal. This group of carbo-hydrates includes the various kinds of sugars, starches, and gums. Carbo-hydrates are always present in a diet. They abound in rice, sago, potato, and bread, and in vegetable matters used as foods.

I need hardly demonstrate to you that milk contains butter. Butter is a mixture of various fats, and as fats are soluble in ether, it is easy to make an ethereal solution of the fat of milk by shaking up milk with ether, after adding a little caustic potash to it and keeping it moderately warm. In this long tube you see a layer of ether holding fat in solution, floating on the top of the fluid. Fat also consists of carbon, hydrogen, and oxygen, but in proportions different from those in which these elements exist in the carbo-hydrates. It contains no nitrogen. Fatty matter must always exist in a diet suitable for sustaining life.

Then we have in milk various mineral or saline substances, such as chloride of sodium or common salt, chloride of potassium (possibly), phosphates of soda and potash, phosphates of lime and magnesia, and the ash always shows traces of iron, although we are not acquainted with the exact condition in which iron exists in milk. Here is some ash of milk prepared for us by the chemists, and it would not be difficult to show you that these salts exist in it. Lastly, milk always contains water as the solvent for all the substances I have mentioned. Examine for a moment this table.

TABLE SHOWING THE COMPOSITION IN 100 PARTS OF VARIOUS MILKS.

	Cow.	Goat.	Mare.	Dog.	Human.	Condensed Swiss.	Condensed English.
Proteid matter chiefly casein	3·34	4·20	2·70	11·70	2·45	10·20	11·84
Fat (butter)	3·53	5·80	2·50	9·72	3·10	9·76	8·30
Carbo-hydrate (sugar)	4·75	4·94	5·50	3·00	6·70	51·02	50·79
Salts . . .	·75	1·00	·50	1·35	·30	2·32	2·00
Water . .	87·63	84·06	88·80	74·23	87·45	26·70	27·07

Milk, then, a typical food, contains proteid or nitrogenous matter in the form of caseinogen,[1] carbo-hydrate in the form of milk-sugar, fat as butter, saline matters, and water. All these

[1] A little albumin is also present.

so-called proximate principles must exist in any dietary that can keep an animal healthy and strong. The reason of this is, that if you examine chemically almost any of the tissues of the animal, you find that they are built up of the same kind of proximate constituents. Suppose a chemist analysed muscle, he would find in it proteid matters in the form of a substance called myosin, along with other albuminous bodies; carbo-hydrates would be represented by glycogen, a kind of animal starch, and by sugar; fats are there also; and if he burnt the muscle, an ash would be left containing the same salts as we found in milk. But the constituents in the food stuffs are not quite the same as those in muscle, and they are therefore subjected to many chemical processes in digestion by which they are first converted into stuffs that exist in the blood. From these stuffs in the blood the muscle builds up its substance. Now all of these stuffs, from the scientific point of view, contain energy in what is spoken of as a potential state, that is to say, it is resting, ready to be set free, ready to do work, and when it is set free it may become, as it does become in the case of the muscle, mechanical energy and heat. This locked-up energy is liberated by one familiar process, that of burning. But burning is oxidation. The elements of the substance to be burnt are torn from each other, and they unite with the oxygen of the air to form simpler bodies. When this occurs, energy appears as heat. Similar phenomena occur in a muscle. The muscle needs oxygen and it needs food stuffs. It builds some of these food stuffs into its own substance, thus always making up for the "tear and wear" that goes on when it works, and it uses some of the others, effecting in some mysterious way chemical changes in them, always splitting them up into simpler bodies.

Let us think of the muscle as a little machine or engine. When it works, it is subjected to tear and wear, like any other machine, and by and by it would become unfit for work. This happens, as we all know, with any machine. The boiler plates of an engine get thinned, the pinions become slack, the wheels do not work so smoothly as at first, and a time comes when the engine becomes only old iron and is unfit for use. But while the same process of tear and wear goes on in a muscle, the muscle, being a living thing, has the power of self-repair. It is always engaged in mending itself, building up so as to make good the waste, and in this way for a long time it is able to work efficiently. The substances needed for building it up are brought to its own door by the blood, many of them ready made, and it takes them into itself and repairs the machinery. These substances for repair are no doubt the proteids, the carbo-hydrates, the fats, saline matters, and water. They all seem to be necessary for the upkeep of the muscle-substance.

But our little machine not only keeps itself in repair, but it can excite chemical changes in certain matters brought to it, and by these changes energy is liberated. There are strong grounds for holding that the carbo-hydrate matters are changed or altered in this way by the action on them of the living muscle-substance. The history of these carbo-hydrates is very wonderful. Entering the body mainly as starch, they are changed into sugar; then they pass into the blood and are carried to the liver; then they are reconverted into the animal starch, glycogen, and stored up in the liver for further use; lastly, the glycogen is again changed into sugar, either in the liver or in the muscles; and in the muscles the sugar is used up, being ultimately decomposed into carbonic acid and water. The splitting up of this sugar or carbo-hydrate in the muscle is, there is every reason to think, the main source of the liberation of the energy in the muscle. This view explains in particular the greatly increased consumption of oxygen and the greatly increased production of carbonic acid following active muscular exertion. At least a part of the carbonic acid is the waste product arising from the decomposition of the carbo-hydrate.

Now if the muscle receives no carbo-hydrate, or an inadequate supply of it, it does not follow that it will stop working. Experiment has shown the contrary. It will still work, using up the fat in the first instance; and if there is an inadequate supply of fat, it actually uses up proteid matter, or, in other words, it uses its own substance. If an animal is caused to work very hard, we do not find an increase in the excretion of the nitrogenous waste matters, as Liebig supposed; but, as I have said, an increase in the non-nitrogenous waste matter, carbonic acid. So long as carbo-hydrates or fats are freely supplied, no increase in nitrogenous waste matters follows prolonged muscular exertion; but if they are withheld, or if the muscular exertion is excessive, then there is an increase in the waste nitrogenous substances. This shows that when the little machine cannot give its output of energy at the expense of carbo-hydrate or of fat, it sacrifices a part of its own framework. The muscle engineer, when he finds himself short of the ordinary fuel, seizes hold of combustible portions of his own engine, as if he were determined at all costs to do the work required of him. This is only a somewhat fanciful analogy, but it gives us an insight into what probably occurs in a muscle.

An engineer is desirous, chiefly for the sake of economy, to get as much effective work as possible out of the engine he constructs. The engine is intended to do work by liberating mechanical energy; but part of the energy appears as heat, and the heat is of no use to the engineer. The engineer knows the amount of energy represented by the fuel he uses, and he can estimate the amount of effective mechanical energy he can get out of it by the best arrangements yet devised. We are told that the best triple expansion steam-engine, with the best arrangement of furnaces and boilers, gives back of effective mechanical energy only about twelve and a half per cent of the total energy in the fuel. This means that of every one hundred parts of energy supplied to the engine only twelve and a half parts are of any use, the remaining eighty-seven and a half parts being lost as heat. The case is worse when we consider an ordinary locomotive, by which only about four per cent of the total energy becomes effective. It is interesting to inquire how the muscle, considered as a little engine, compares with the best steam-engine.

So long ago as 1869, Professor Fick of Würtzburg stated that the amount of energy transformed into mechanical energy by a muscle was about thirty-three per cent of the energy in the food stuffs. In 1878, he announced that more accurate experiments had obliged him to reduce the estimate to twenty-five per cent. Fick's experiments were made on the isolated muscles of frogs, and to some extent were vitiated by the conditions in which the muscles were examined.

To illustrate what is meant by the work of a muscle, I show you here an interesting experi-

FIG. 63.—The work-gatherer of Fick. It may be called a muscle-winch or muscle-capstan.

ment carried out by an ingenious instrument devised by Professor Fick, and which he calls the *Arbeitssammler*, the work-gatherer. It is a little windlass or capstan, which you see is turned by a frog's muscle placed above it, and

Fig. 64.—Arrangement of the muscle-capstan. The shadow was projected on the screen. The muscle was placed above the capstan and the nerve was irritated at regular intervals, say of one minute, by tetanising shocks lasting only for a short time, say three seconds. By means of the pulley placed below the stand bearing the capstan, the lifting of the weight (made of cork) with each contraction of the muscle was so large as to be readily seen.

which I stimulate at regular intervals of time. You observe a little catch on the edge of the wheel, which keeps the wheel from going in the opposite way during the relaxation of the muscle. Consequently the muscle, as you see, winds up the weight. Now if we multiply the weight by the height through which the muscle has lifted it we get a measure of the work done. We speak of a foot pound, that is one pound weight lifted one foot in height, or we speak of a kilogrammetre, that is one kilogram lifted one metre in height. In like manner, in estimating the work of a muscle, we use the phrase gramme-millimetre, that is one gramme lifted one millimetre in height, or about fifteen grains lifted the one-twenty-fifth of an inch. You observe how easy it is to get a notion of what musclework means by the use of this beautiful instrument.

It has been found that the work actually obtained from a frog's muscle may be stated as follows: one gramme of muscle (that is about fifteen grains) will yield four gramme-metres of work. A gramme-metre is one gramme lifted one metre (a little over three feet). Four gramme-metres represent, then, fifteen grains lifted a height of twelve feet. This may seem a small amount of work, and it would be so if, in doing it, the gramme of muscle disappeared; but only an infinitesimal part of the muscle-substance is used up in the experiment, and the most careful weighing would probably fail in detecting the loss. Perhaps not more than the one-thousandth part of the fifteen grains of muscle has been used to lift that weight twelve feet high.

Chauveau, one of the greatest of living French physiologists, has recently reinvestigated the question by ingenious experiments on the muscles of living men in normal conditions. These experiments oblige Chauveau to reduce Fick's estimate and to give the total effective energy as only from twelve to fifteen per cent. Taking the total mechanical energy of a man instead of a muscle, some recent calculations of my own show an output of mechanical energy as a little over seventeen per cent.

It is evident, therefore, that, considered as an engine, a muscle is not much better as a transformer of energy than the best steam-engine now constructed, while it is inferior to certain gas-engines which are said to return as much as twenty per cent in the form of effective mechanical energy. The muscle, however, has this great advantage over any engine, that the heat it produces supplies one of the conditions of its very existence, the maintenance of a certain uniform temperature. Muscle-substance will only work within a limited range of temperature, and as the body is constantly losing heat by radiation, conduction, evaporation of sweat, and by other means, heat must be supplied. This comes mainly from the muscles. It would not be correct to say, however, that one of the final purposes of a muscle is to produce heat. It is not a heat-producing machine. The primary function of a muscle is to contract, and thus to do effective work; but by producing heat at the same time it becomes possible for the muscle-substance to do its work in the best possible conditions. This is only another illustration of that wise economy that we see in most, if not in all, of the arrangements of nature.

After considering the points discussed in this lecture, you will readily understand how it is that a muscle becomes tired. It becomes fatigued after continuous work. Fatigue means a diminished power of work. Up to a certain point, the substances produced in a working muscle are got rid of as quickly as they are formed and new materials are supplied for the repair of the muscle. There is thus a balance between the two processes. But if a muscle is made to work for a long period, or if it is excited to very frequent contractions, the waste products gather or accumulate in the muscle, and sufficient time is not allowed for the supply of reparative material. Muscle, like most other tissues, is richly supplied by a special set of vessels, of very minute size, called the lymphatics, which are for the purpose of draining away the excess of nutrient matter that has oozed out of the vessels, along with the waste matters that have been formed. If the waste stuffs are produced too quickly—such stuffs as carbonic acid, acid phosphate of potash, and the stuffs of a nitrogenous nature, such as kreatin and other bodies found in extracts of meat—the muscle becomes fatigued; it consumes less oxygen and produces a smaller amount of waste products.

It is very instructive to watch how a muscle behaves as it becomes tired. I have fitted up this beautiful experiment, as devised by Marey, to show you this. The pithed frog (entirely devoid of sensation) lies on this cork plate, and a thread from the tendon of

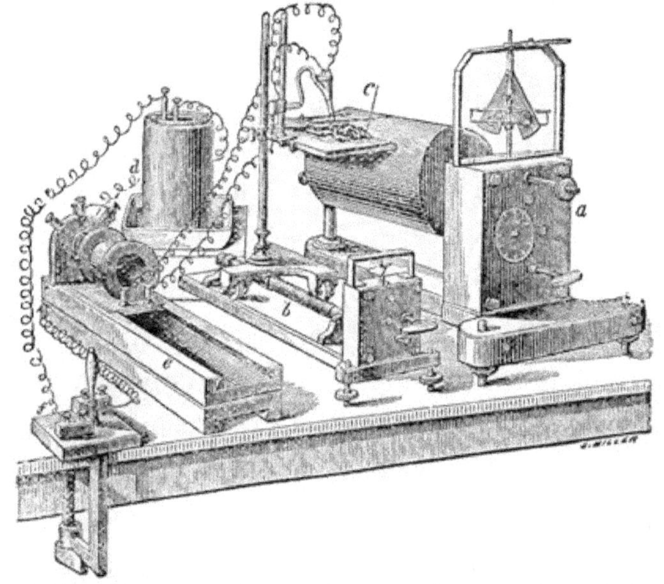

FIG. 65.—Arrangement of apparatus for the demonstration of fatigue. *a*, recording cylinder ; *b*, railroad carrying the myograph, *c* ; *d* galvanic element ; *e*, induction coil ; *f*, key.

the gastrocnemius passes to a light lever writing on the surface of this drum. The cork plate, bearing the frog, is on a stand that moves by clockwork from left to right, so that the plate moves a little to the right during one revolution of the drum. Thus the tracings are kept from blurring, each successive curve being a little to the right, or, to put it in another way, a close threaded spiral is described round the drum. Now we shall irritate the nerve by an opening and a closing shock from the secondary coil of our induction machine, each shock coming always at the end of an equal interval of time. This we can arrange by attaching a wire to the axle of the cylinder, so that it stands out at right angles to the axle, and as the axle revolves, the wire dips into this trough containing mercury, thus closing the current of the primary coil of the induction machine, and the next instant the wire comes out of the mercury, thus opening the current of the primary coil. As you know, this will secure a shock from the secondary coil each time the wire dips into and comes out of the mercury. The current is led from the battery, through the primary coil, thence to the cylinder, thence through the wire into the mercury, thence back to the battery. Now notice the beautiful curve. It shows, first, the gradual lengthening of the period of latent stimulation, as indicated by each contraction beginning a little later than the one before it. You observe the gradual slope of the line joining the beginnings of the successive curves. Second, we find that at the beginning of the experiment the amount or amplitude of each contraction

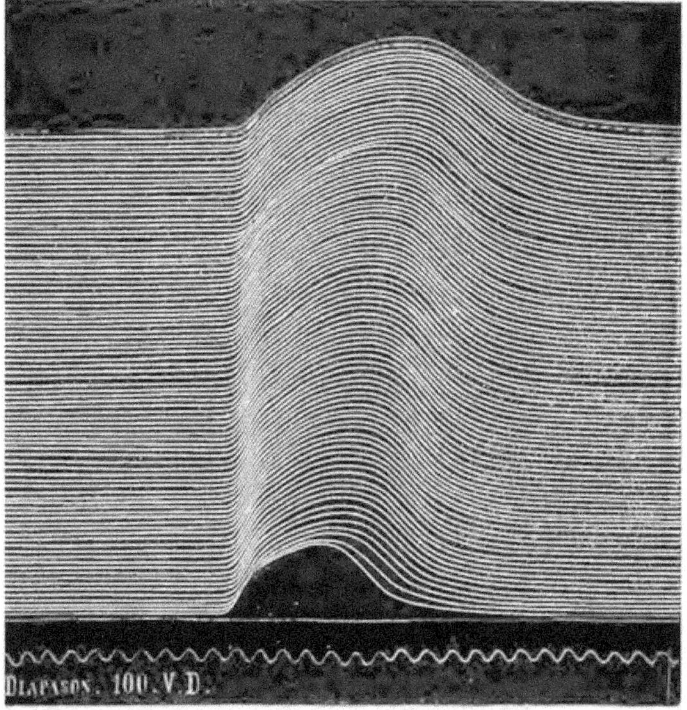

Fig. 66.—Consecutive tracings of the contractions of the gastrocnemius muscle of a frog, showing the effects of fatigue. Chronograph, 100 vibrations per second. Marey.

slightly increases, or in other words, by successive stimulations, the muscle gets into good working trim. In the curve in the diagram this increase is seen, as the curves at the top of the diagram are higher than those at the foot. Fatigue, so far as amplitude of contraction is concerned, has scarcely begun. By and by, however, as the muscle becomes fatigued, the amount of contraction diminishes, until the muscle does not contract at all, but the duration of the contraction increases throughout the whole contraction. The muscle is gradually losing time in doing its work. When does it lose time? In contracting or in relaxing? You observe the slope of the successive curves; the way in which they open out, as seen if you study each from the bottom to the top of the diagram, shows that it loses time during relaxation. During fatigue a muscle after contracting returns more slowly to its original length.[1]

These results are consistent with our experience. After a thirty mile walk, we feel unwilling to take each step; it is only by a strong effort of the will that we force the muscles to contract. Like jaded horses, they require the whip and spur. The muscular contractions required for each step, however,

[1] A beautiful series of curves illustrating fatigue taken by the railway myograph was also shown by placing the glass plate in the lantern.

are not shorter in duration. When the muscles do respond they contract as usual, perhaps not to so great an extent, but then they relax slowly, and we wearily drag our limbs onwards. I do not say that fatigue is entirely in the muscles. They communicate with headquarters and they telegraph their wearied condition to the executive, and the executive also becomes tired, partly by receiving these messages from the muscles, and partly by having to stimulate the flagging muscles to a much greater extent than when they are fresh and active. Using the whip and spur may weary the rider while they stimulate the steed.

There is one other peculiarity of the muscle, considered as a machine, that distinguishes it, to some extent at least, from all other machines. You are all aware that in a great factory containing complicated machinery it does not pay to allow the machinery to stand still, even suppose it may be working at a daily loss to the manufacturer. It may so happen, and unfortunately it not unfrequently happens, that bad trade compels a manufacturer to produce his articles at a loss, and you might think it would be prudent on his part to stop his machinery. But this would probably be an unwise step, both because he might lose the market, and still more because his machinery would deteriorate by standing idle. The manufacturer therefore prefers to hear the whirr and roar of his machines, although he may feel that they are working at a dead loss to him. He does not expect that the work will improve his machines, but he is sure that it will keep them in good condition.

The muscle-engine, in like manner, deteriorates if it is allowed to stand idle. Sometimes this occurs in disease. In paralysis the limb cannot move, and the inactive muscles waste, become thin and flaccid, and undergo curious molecular changes, converting the muscular matter into fatty-like particles. The wise physician, however, knows this, and, like the manufacturer, he keeps the machinery running. He does this by stimulating the muscles by electricity and causing them to contract. The electricity supplies the stimulus that the nerves cannot give, and the physician keeps up the strength of the muscles and hopes for better times.

We often, however, fall into the bad habit of allowing our muscles to be inactive, and the result is they become weak and attenuated. Exercise is needed. Run the machinery and you obtain the wonderful result that the living machine improves in strength and size. The mere act of making the muscle work develops its powers. It grows stronger and thicker, and it works with greater precision and effectiveness. Hence the value of athletic exercises, if carefully carried out. They should not, however, fall on one muscle or one set of muscles exclusively. If they do, these muscles, instead of being benefited, suffer from fatigue. Athletic exercises should be carefully graduated and selected, so as to employ different groups of muscles, stimulating and developing each without unduly exhausting any one group. It is often forgotten, I think, that this is best accomplished by natural movements. The lower animals—for example, take a cat, in which the muscular arrangements are admirably developed—do not require to go to gymnasia for graduated athletic exercises. They run, and leap, and move in the almost unconscious enjoyment (if I may use a phrase that appears self-contradictory), of their physical organisation, and in doing so they develop each part of it. In like manner, the human being should, at least in early life, run, and leap, and play, and in more advanced life a good long daily walk will supply all that is necessary. There is a good deal of philosophy in this, as in many other common things.

LECTURE VI

Electrical phenomena of muscle—Current of resting muscle—Current of acting muscle—Negative variation—Animal electricity—Electric fishes—Resemblance of electric organ to muscle—Relation of muscular motion to nervous system—Conclusion.

To-day I wish, in the first place, to demonstrate to you certain properties of muscle to which I have not alluded. I will show you that muscle has electrical properties. Let us examine some of the facts that led to the discovery of animal electricity, a discovery of the most momentous consequences to the human race.

At the back of the lecture-theatre we have placed a very sensitive galvanometer, specially constructed for the kind of work to which it is now to be put. It belongs to Professor Dewar, and was presented to him by the late Mr. Warren de la Rue, whose zeal in the cause of science is well known to all connected with this Institution. Professor Dewar has kindly

FIG. 67.—Wiedemann's galvanometer, much employed, especially in Continental schools, by physiologists. The instrument used in the lecture was a Thomson's (Lord Kelvin) galvanometer, and it is shown in Fig. 58, *b*. Wiedemann's instrument I have often used with great advantage for class demonstration. The two outer coils are of low, and the two inner coils are of high, resistance. The low resistance coils are used for thermal currents, and the high resistance coils for the currents of living tissues. A ring magnet is suspended by a long filament of silk, and hangs in a copper box in the centre of the coils of wire. On the rod carrying the ring-magnet, and in the box above the coils, we have a mirror, which reflects a beam of light on to the scale.

placed it at my service, and I wish to say in a

word how much I feel indebted to him for the great interest he has taken in these lectures. This instrument has a resistance of no less than 86,000 ohms, and it is one of the most sensitive of its kind.

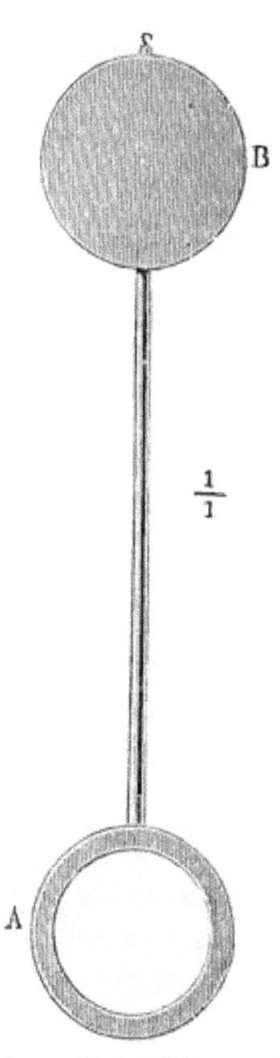

Fig. 68.—Mirror and ring-magnet of Wiedemann's galvanometer. B, mirror; A, ring-magnet.

Now if I brought the wires of this instrument into direct contact with the muscle, we would probably get a current; but that would be no proof that the current really came from the muscle. The copper wires come into contact with the moist surface of the muscle, and this contact would at once, by chemical action, generate a current of electricity. For example, to show you how easily currents can be produced and detected by an instrument of this kind, observe that when I touch the wires the spot of light at once moves. We must have some means, therefore, of leading off from the muscle any current that may be produced by it, without generating currents in the apparatus we employ. The question arises—what contact can be made with a muscle without generating currents at the surface of contact?

After much labour, Professor E. du Bois Reymond of Berlin invented the proper appliances. Let me mention in passing the name of du Bois Reymond with much respect. He has not only laid the foundations, but he has built much of the superstructure of our knowledge of electro-physiology; he, more than most men, has investigated the hidden processes in muscles and nerves, devising and even constructing, in the first instance, with his own hands, much of the apparatus now employed in such investigations; and it is interesting to know that he demonstrated, in 1855, many of his discoveries at a famous lecture given in this Institution. Well, du Bois Reymond found that zinc troughs, carefully rubbed over with mercury, or, as it is termed, amalgamated, and filled with a saturated solution of sulphate of zinc, fulfilled the conditions. Into these troughs we place pads of white blotting-paper (Swedish filter-paper). But if we laid the muscle on these pads, the sulphate of zinc solution would irritate the muscle, and that, we shall see, produces new phenomena. We wish to examine the electrical properties of the muscle at rest. To protect the muscle,

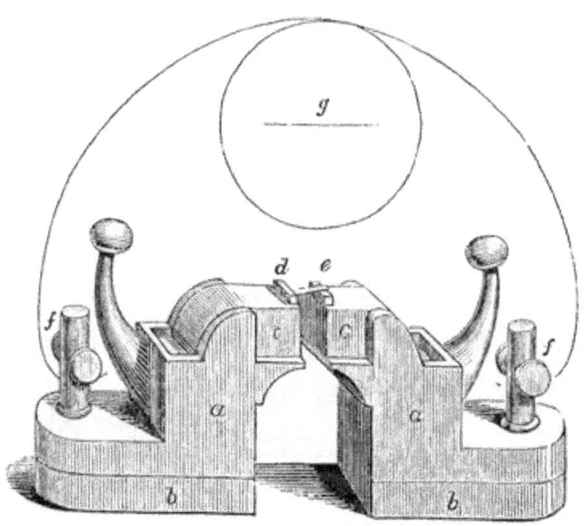

FIG. 69.—Arrangement of du Bois Reymond's troughs. *a a*, troughs of zinc; *b*, vulcanite plates for insulation; *c c*, paper pads; *d e*, clay pads; *f f*, connections for wires leading to galvanometer, *g*.

therefore, we place on the pads of blotting-paper little bits of moist clay: sculptor's clay, moistened with saliva, is usually employed. Here are the troughs just as they are used by their inventor. There are many other forms of these electrodes, or instruments for leading off the electricity, adapted for special use; but

I prefer to use the original form, as it is the one I have employed for years, and with which I am familiar.

Our troughs or electrodes are connected with the galvanometer, a key being interposed in the circuit. Notice where the spot of light is at present. I shut the key, and you see there is a very slight movement of the spot of light, showing that the troughs are already producing a certain amount of current. This arises from the fact that somewhere a slight chemical change is going on, quite sufficient to generate a feeble current. Let us remember, however, that the current of the troughs makes the spot of light go to the right. Now we know, or at all events the electricians tell us, that copper is positive to zinc, which is said to be negative; or, in other words, if we placed a copper and zinc plate in a fluid acting on the zinc, a current would travel out by the copper, and if it had a completed circuit, it would return to the zinc. Well, observe when I touch with one hand the copper wire connected with the trough on my right, and with the other hand the zinc trough on my left, the spot of light moves to the right; but when

I touch the copper wire on the left and the zinc trough on my right the spot travels to the left; that is to say, when the right-hand trough is positive the light moves to the right, and when the left-hand trough is positive it moves to the left. We must keep this in mind.

I now take this muscle—the gastrocnemius of a frog—and, with a sharp pair of scissors, cut it clean across the fibres. Observe that I lay the muscle on the clay pads, so that the surface of the muscle touches the pad on the right, and the cut surface, that is, the transverse section, touches the pad on the left. I close the key so as to allow any current that may exist to flow to the galvanometer, and you see that at once the spot of light swings to the right. Observe also that the spot of light keeps to the right, showing that a current is flowing through the wire of the galvanometer. You will remember, however, that without the muscle there was a small amount of movement to the right, but the movement you now see is much greater, and cannot be due to the cause that produced the previous small movement. I open the key so as to break the circuit, and at once the spot of light slowly sails back to the original point. You will recollect that when the right-hand trough was positive, the spot travelled to the right, as we have just seen.

Pursuing the experiment farther, I now pick up the muscle with the forceps and place it again on the pads, but in a reversed position; that is to say, the surface now touches the pad on the left, while the section is in contact with the pad on the right. I again close the key, and you now see the spot passes to the left and takes up a position on that side. Open the key, and it again sails back. Remember, once more, that the spot should come to the left when the left pad is positive, as has occurred in this last experiment. These two experiments clearly prove, first, that a muscle at rest gives a current; and, second, that this current travels through the galvanometer circuit from the surface of the muscle to the transverse section. The surface of the muscle is thus positive to the transverse section; or you may get a clearer notion of the statement by supposing the bit of muscle to be a little galvanic element or battery. In that case, the surface of the muscle is the positive pole, and the transverse section is the negative pole. This then is a demonstration of the electrical condition of the muscle at rest.

Now let us go a little farther. Leaving the muscle on the pads, with the longitudinal surface touching the pad on the left, and the transverse section touching the pad on the right, you notice the spot of light has taken up a position well towards the left. The spot will slowly move to the right as the muscle dies, but we need not wait for this. I need hardly say that a dead muscle gives no current. What would happen if we made the muscle contract? Would the contraction increase or diminish the current, or would it have no effect? To answer this question, I shall ask Mr. Brodie to lay the nerve still attached to the muscle over the wires coming from the induction coil, and, as he has a key in the circuit, I ask him by closing it to throw the muscle into tetanus while it remains on the pads. Observe once more the position of the spot of light. He now tetanises the muscle, and you see the spot immediately travels to the right, indicating either that the current is less during contraction than it was before, or that a current is now flowing in the opposite direction. This swing backwards will carry the galvanometer needle to the point from which it started, or even to the opposite side. It is known technically as the negative variation of the muscle current; and careful experiment has shown that it is not due to a mere diminution of the current from the muscle while the muscle is at rest, but that it is really a current in the opposite direction; that is to say, when a muscle having a cross section contracts, the surface of the cross section becomes positive instead of negative, and the longitudinal surface becomes negative instead of positive. As a current always travels from positive to negative, it follows that the current in a contracting muscle travels in the opposite direction from its course in a resting muscle.

Other tissues than muscle show so-called resting currents, and when the tissue acts or responds to a stimulus, there are strong indications that a negative variation current is produced. Thus a nerve shows the same phenomena as a muscle. Currents have also been observed in glands.

One of the most striking demonstrations of these currents is the one I shall now show you. The frog's heart beats for a long time after the death of the animal. Here is a little heart still beating. I cut off with the scissors a small portion of the apex, and I place the heart on the pads, so that the surface touches one pad and the cut apex the other. I now close the key, and you observe a swing of the galvanometer with each beat of the heart. You observe the spot of light is unsteady; it swings to the right with the current of the heart at rest, and to the left with the current of the heart when it contracts. The latter is the negative variation current. Sometimes we even get a double swing of the galvanometer. These beautiful results are shown in another way by my friend Dr. Augustus Waller. He uses an instrument called a capillary electrometer, invented by Lippmann, and he can demonstrate the electrical variations of the human heart.

The current of the cut muscle at rest is not of so much importance as the current of the same muscle in action. There are strong reasons for holding that the resting current is due to the death or dying of the layer of muscle-substance laid bare by the transverse section. Dying muscle-substance, it would appear, becomes negative to living substance. But the negative variation current, or the action current, as we may well term it, indicates changes happening in the muscle, changes that are somehow connected with the phenomena of contraction. It may be detected by special methods without even injuring the muscle.

It was supposed, until recently, that the negative variation occurred solely in the period of latent stimulation, a period perhaps as short as the one-two-hundredth of a second; but by special photographic methods of exquisite delicacy, Professor Burdon Sanderson has shown that this is not the case, and that it continues into the time of the contraction of the muscle. Nor have we any definite proof that these electrical changes are dependent on the chemical changes discussed in last lecture, although it is highly probable that the two are intimately connected.

May not the negative variation change in one set of muscular fibres do something in the way of irritating adjacent fibres? There is an old experiment devised by Matteucci that favours this view. I have two of the usual nerve and muscle preparations connected with these two telegraphs. Call this one A and the other B. I stretch the nerve of A over B, and the nerve of B is placed on the wires of the induction coil. I now irritate the nerve of B, and of course its muscle B contracts;

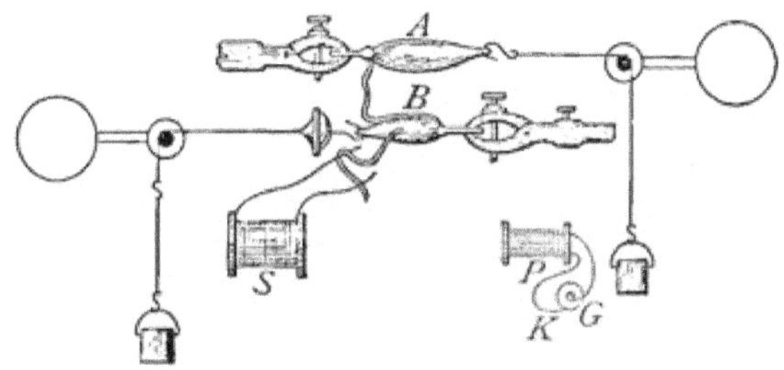

FIG. 70.—Arrangement to show Matteucci's induced contraction. A, muscle, the nerve of which is placed on B; G, galvanic element; K, key; P primary and S secondary coil of induction machine.

but you observe the muscle A also contracts. We explain this by supposing that the negative variation change in the muscle B is sufficient to stimulate the nerve of A, and therefore A contracts as well as B. Matteucci, in a similar way, found that a number of muscles might be thus connected, the nerve of the one lying on the muscle preceding it, so that when the nerve of the first member of the chain was stimulated all the muscles contracted.

If this be the case, may we not suppose, as was suggested by Professor Kühne of Heidelberg some time ago, that those parts of certain muscles which are destitute of nerves may be

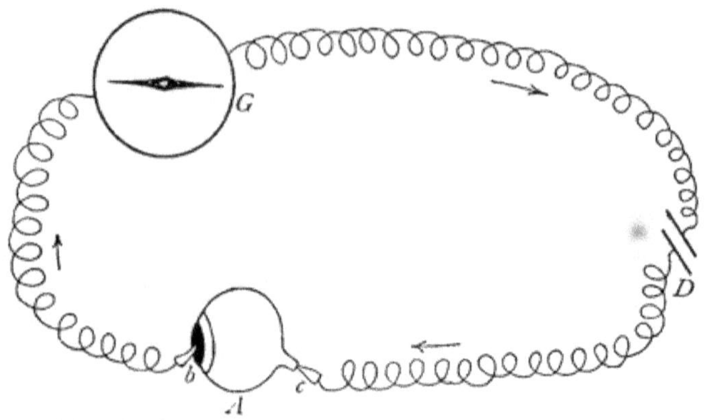

Fig. 71.—Diagram showing arrangement of apparatus used in demonstrating the action of light on the retina of a frog's eye. A, the eye, having one electrode, *b*, touching the centre of the cornea, and the other, *c*, touching the transverse section of the optic nerve; G, galvanometer; D, key. The arrows show the direction of the current.

thrown into action by the stimulus of the negative change happening in adjacent portions of the muscle supplied with nerves? I think this is highly probable.

I shall now endeavour to show you the electrical change produced by the action of light on the frog's eye, a subject on which

Professor Dewar and I worked nearly twenty years ago. The eye of a frog has been carefully dissected out (the animal of course being dead, although the tissues of its eye still live), and it has been placed on the clay pads so that one pad touches the back of the eye while the other is in contact with the front of the eye—the cornea. We now place the eye in darkness by covering the troughs over with a bandbox, in which, however, we have left a small window which we can open and shut at pleasure, and the position of the window is such that if we place a light before it, the light will shine on the cornea of the little eye. Now I shut the key so as to allow any current that there may be to flow to the galvanometer. You see at once there is a very considerable current. That is the resting current of the eye in the dark. Mr. Brodie will now allow light to fall on the eye. You see at once the spot of light on the scale moves and indicates an increase in the first current. As light continues to act the current begins to diminish; but I now ask Mr. Brodie to take away the light and leave the eye in darkness. You see the moment the light was removed that the current again suddenly increased and then fell to a point lower than it had hitherto been. We repeat the experiment and you observe we get the same results. A sudden influx of light causes an increase in the current of the living eye; under the continued action of light the retina becomes fatigued, and, when light is taken off, there is another increase and then a great falling off. You see how sensitive the eye is —even a flash or striking a match near it produces the effect. If I let the light pass through this bit of red glass we get an effect almost equal to what we got with yellow light from the taper; but if it passes through this dark blue glass, you notice the effect is much less. This is perhaps the most delicate experiment in the range of physiological science.

We have now examined the muscle-current, the nerve-current, the heart-current, and the eye-current. Let me next endeavour to show you the man-current. I have here two flat vulcanite troughs into which we have poured a three-quarters per cent solution of common salt. Mr. Brodie has placed one zinc trough by the side of each flat trough, and he has dipped the points of the clay pads into the salt solution. We first of all connect the vulcanite troughs together by a bit of wet blotting-paper, so as to put them in circuit, and I now close the key. If any current came from the troughs themselves, we would see a movement of the spot of light on our scale. You observe there is scarcely any movement. I now place my hands in the troughs, laying them in the salt solution. At first there is a considerable movement of the spot of light on the scale, but it soon comes to rest. If, however, I contract the muscles of my right arm, the spot moves to the right with a great swing, and if I contract the muscles of my left arm the swing is in the opposite direction. You observe by alternately contracting the muscles of the right and of the left arm I can cause the spot to swing in either direction. Notice how sensitive the effect is, even when I place only a forefinger in each vessel containing the salt solution. This is the man-current. It may be explained by supposing that when the muscles of both sides of the body are at rest there is no difference of electrical potential between the one side and the other side; but if the muscles of one side, say of one arm, are contracted, this at once produces a disturbance, and a current flows through the galvanometer. It is not an entirely satisfactory explanation, but it is the best that can at present be given.

I shall not discuss the theories that have been propounded to explain these remarkable phenomena, the investigation of which clearly demonstrates the existence of a true animal electricity. In 1791, Galvani, who was Professor of Anatomy and Physiology in Bologna, first announced that electricity, to use his own phrase, was secreted in, or originated from, the animal tissues. The great controversy that then arose, more especially between Galvani and Volta, who was Professor of Natural Philosophy in Pavia, led to the invention of the Voltaic pile in 1799, and still more to the discovery of the production of electric currents by the contact of dissimilar metals, more especially when one is acted on chemically by certain fluids. For a long time the brilliancy of the results flowing from investigations into Voltaic electricity threw the discoveries of Galvani into the shade; but by and by, as methods of observation became more refined, it was found that there is in truth an animal electricity, and that Galvani was right in many of his views.

One branch of science often helps another. By the discovery of the influence of a current of electricity on a magnetic needle, made, in 1820, by Oersted, the galvanometer or multiplier became possible. Nobili, about 1825, constructed such an instrument for physiological purposes, and again demonstrated the muscle-current. In 1837, Matteucci enriched the subject by many beautiful investigations, and, in 1841, du Bois Reymond took it up with rare enthusiasm, and from that year to the present year has laboured on it with much success. One feels, after reading du Bois Reymond's monographs, that he has left little for the gleaners in this great harvest.

How striking is it, my young friends, that the splendid results of modern electrical science, with which we are familiar every day, flowed, in no small degree, from the first physiological experiments of Galvani. Electric lighting, the application of electricity to the construction of motors, and the thousand ways in which this mysterious thing is becoming the servant of man, sprang from discoveries that date from the time when the Italian philosopher noticed the twitches of a frog's leg near his electrical machine. Truly there is nothing small and insignificant in science. An observation of a phenomenon obscure in character and not striking to the senses may be the key by which we open new stores of knowledge.

It has often been remarked that many of man's inventions have their counterpart in nature. In the subject we are discussing we have a proof of the truth of the remark. In nature we find living electrical machines. When one considers how potent electricity is, a kind of "vril" (to use the word coined by the author of *The Coming Race*), armed with which a being might become a terrible antagonist, it is remarkable that electric organs have as yet been found only in a few species of fishes. No doubt the explanation may be offered that the conditions of the environment of living things are not favourable to the general evolution of electric organs, and while this explanation is probably true, it does not lessen, I think, the sense of astonishment. Some fifty fishes possess electrical organs, and of these only five or six have been investigated. These remarkable creatures are of interest to us at present because the electrical

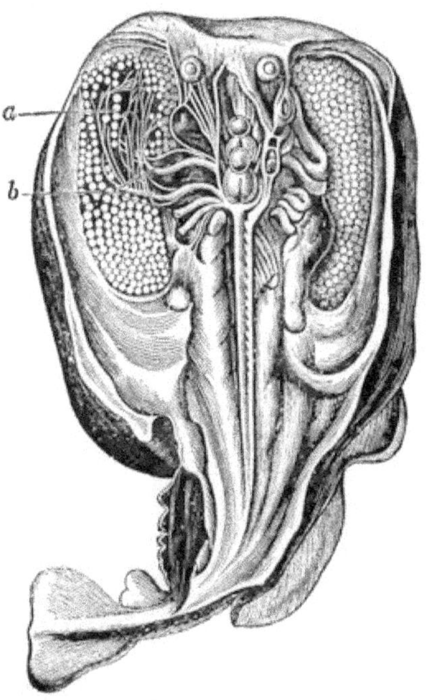

Fig. 72.—*Torpedo Galvani*, showing the prisms of the electric organ as seen from the dorsal surface. Each organ contains about 800 prisms, and each prism is divided by delicate membranous plates, separated from each other by a jelly-like fluid. Each prism has about 600 plates, and as there are 800 prisms in each electric organ, the organ contains about half a million electric plates, each of which is supplied by a nerve filament. The figure shows the large nerve trunks, *b*, ending in the smaller nerves, *a*, distributed to the prisms. See next figure.

organs of many of them are, in a sense, modified muscular structures. Thus in the ray called the torpedo (*Torpedo Galvani*) of the Mediterranean the electric organ takes the place of the outer gill muscles of the fifth gill arch. In ordinary rays, and their distant cousins the sharks, these muscles are powerful organs for moving the lower jaw, but in the torpedo for

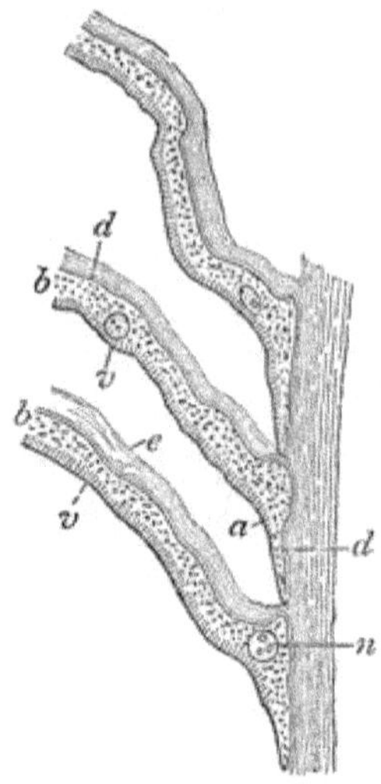

FIG. 73.—Attachment of the electric plates of torpedo to the sheath of the prism. *d*, sheath of prism; *v*, ventral or nervous layer or plate; *d* (to the left), dorsal plate; *c*, fine layer of connective tissue; *b*, intermediate layer; *n*, nuclei of this layer; *a*, a portion reflected from the plate.

these muscles electric organs have been substituted. At an early stage in the development of the torpedo the tissue of the electric organ is like that of an embryonic muscle, showing numerous nuclei, and even a distinct longitudinal and a more faint transverse striation may be seen. Somewhat later, the striations, both longitudinal and transverse, disappear, the nuclei become larger and more numerous, and

FIG. 74.—The electric eel, *Gymnotus electricus.*

the disc-like arrangement of plates begins to appear. The process goes on until the slight resemblance to muscle is entirely lost. I show you here a specimen of a large torpedo from the Hunterian Museum of the University of Glasgow, and in this other jar you see the electric organ dissected out of the specimen. This is an interesting preparation. It is very old, and may have been put up by the hands of John Hunter, the brother of William, who founded the museum in my own university.

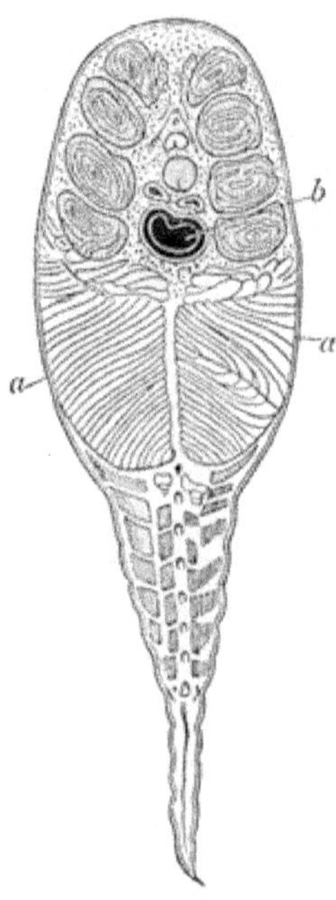

FIG. 75.—Section of the body of gymnotus, showing the position of electric organs. *a a*, electric organ. Above the organs on each side observe the masses of muscle. *b*, swimming bladder.

Again, in the more formidable electric eel (*Gymnotus electricus*) of the region of the Orinoco, in South America, we find huge electric organs running almost from head to tail, which occupy the same positions as are filled by muscles of eels of allied species, and a study of their development shows that they originate from the same kind of substance.

Several species of skates from our own seas have an electric organ in the tail, a fusiform body, about half way up the tail of the fish, in contact with the skin, and partly enveloped in a well-known muscle—the sacro-lumbalis muscle. This organ shows a disc-like structure, somewhat similar to that found in the electrical organs of the torpedo and gymnotus, but more resembling the latter than the former. There can be no doubt again that we may view the electrical organ of the skate as a modified muscular apparatus.

I shall now ask Mr. Brodie to show you sections of these organs by the electric microscope. [This was done, and a demonstration was given, which is partially illustrated by the Figs. 73, 76, 77, 80, and 81.]

But nature shows often a remarkable power of modifying different parts for the same purpose, or of similar parts for different purposes. This is seen both in plant and animal life. Thus the tendrils by which a plant clings to other structures may be modified leaves, stipules, or branches; and, on the other hand, similar parts in many of the

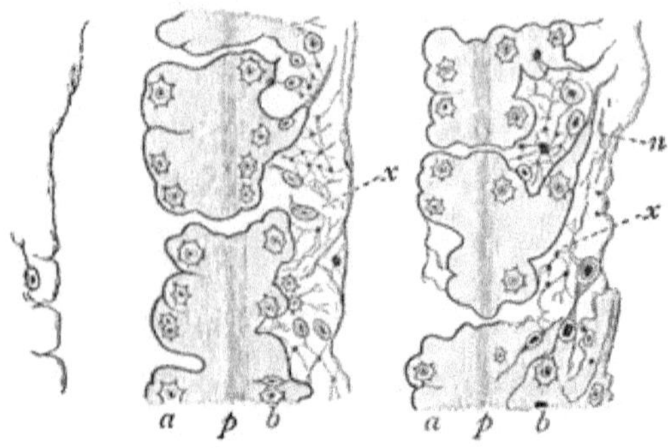

FIG. 76.—Portion of electric organ of gymnotus magnified 400 diameters. *p*, Pacini's line; *a a*, anterior papillæ; *b b*, posterior papillæ, sometimes called thorn papillæ. They contain cells that move (amœboid movements). *n*, nerve-fibres entering papillæ; *x x*, connective tissue. The plates sometimes cleave transversely along Pacini's line, a division analogous to the cleavage of muscle into Bowman's discs.

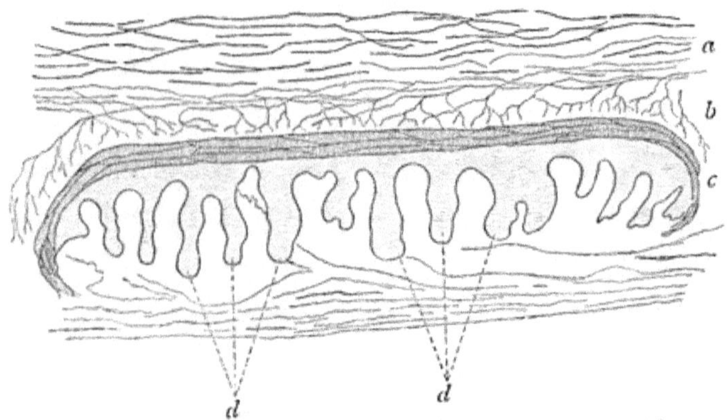

FIG. 77.—Semi-diagrammatic view of a disc from the electric organ of the skate (*Raia batis*). *a a*, connective tissue with capillaries; *b*, nerve layer, nerve endings branching; *c*, striated layer; *d d*, processes of transparent structureless material containing numerous nuclei corresponding to the thorn papillæ of gymnotus.

crustacea (crabs, lobsters, etc.) may become

either gills (organs for breathing) or feet (organs for locomotion). We might expect, then, that electrical organs might be found that were not muscular in their origin.

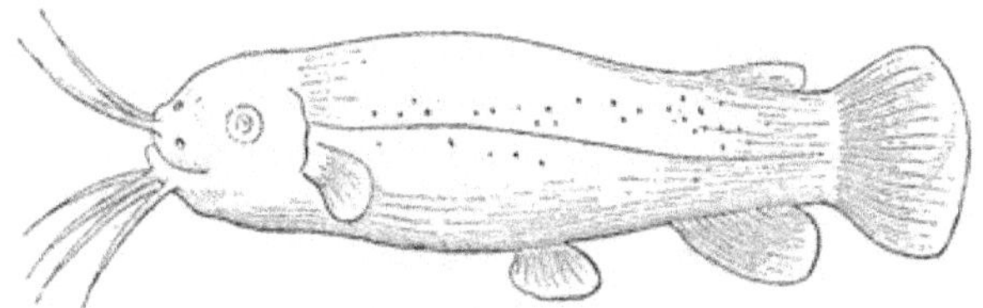

Fig. 78.—The Ra"ásh or thunderer fish of the Arabs. *Malapterurus electricus, Var. affinis.*

Accordingly we find that the electrical organ of the raasch or thunderer fish of the Arabs (*Malapterurus electricus*), an inhabitant of the Nile, is not muscular, but is a modification of peculiar glandular structures found below the skin of allied species.

Electrical organs, in their physiological behaviour, present many striking resemblances to muscle. Thus they are all richly supplied with nerves. When the nerve is irritated the electrical organ discharges electricity, not as a current, but in a number of short sudden shocks, like the quivers of a muscle in tetanus. The battery, however, does not go off at once. There is a latent period preceding the discharge. The electrical organ is connected with the central nervous system, the nerves

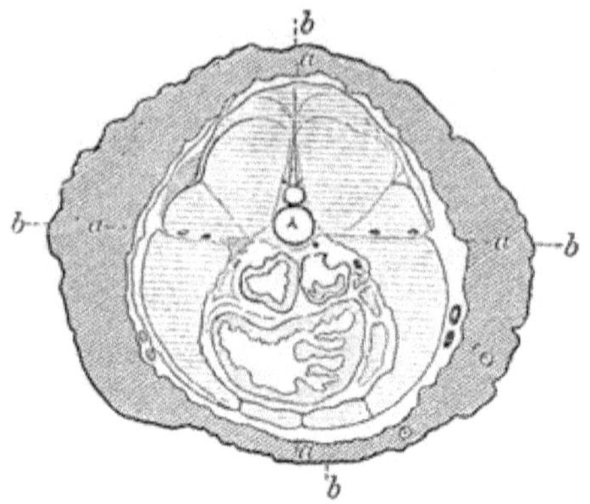

FIG. 79.—Section of malapterurus, showing the electric organ surrounding the body between *a* and *b*.

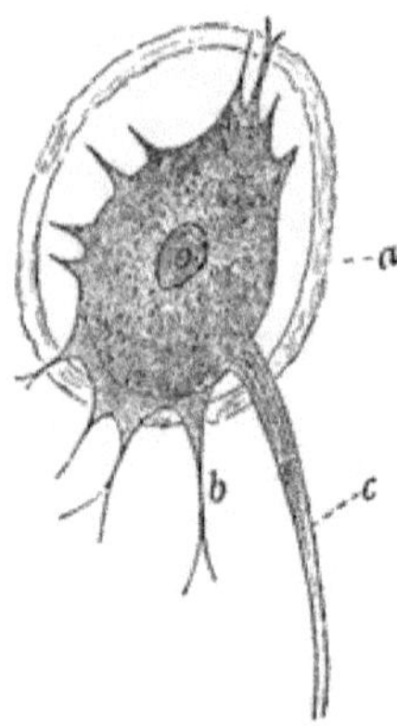

FIG. 80.—Electric cell from the middle region of the spinal cord of gymnotus, magnified 314 diameters. *a*, sheath of neuroglia, the peculiar connective tissue found in the central nervous organs; *b*, smaller nerve processes or fibres, forming a network with those of adjoining cells; *c*, chief nerve process passing into the axis-cylinder of a nerve-fibre. The nerve-fibre ends in an electric plate, as shown in Fig. 76, *n*.

springing from special nerve-cells, so that it

is under the control of the will; but, at the same time, it may be excited to discharge by drugs that act on the nervous centres. Thus strychnia throws the muscles of an animal into terrible tetanic convulsions by acting on its nervous centres; but it causes a torpedo to discharge a quick succession of shocks till the creature is exhausted. Again, an electric organ shows fatigue, and it needs time to rest and recover. Lastly, the organ is the seat of chemical changes.

There are many other phenomena that time will not permit me to mention. If we could get a supply of live torpedoes or of electric eels, and have some means of keeping them alive, I can conceive a course of Christmas lectures of surpassing interest; or we might get over the difficulty of having the live animals in the lecture-room by delivering the lectures in Cairo instead of in London, where no doubt the thunderer would be willing to show his powers. May we, however, hazard an explanation of the nature of these organs and of their relation to muscle and gland? In the case of torpedo, gymnotus, and the skate, the nerve ending is evidently analogous to a motor end-plate in muscle, and in malapterurus to the terminations of nerves in the cells of glands. The molecular disturbance transmitted along a nerve causes changes in its end organ, and these are propagated to the surrounding substance. These changes are associated with a change of potential, and the part becomes negative. A wave of negativity passes through the organ, and there may be a result, the nature of which will depend on the kind of organ in which the change may take place. If it be a muscle, the chief expression of the change is a variation in form or contraction; if it be a gland-cell, the change is the formation, or disintegration, or modification of certain matters of the secretion; and if it be an electrical organ, it is an electrical discharge. In all three, however, similar phenomena occur, but to varying amounts. Thus, call contraction *a*, electrical phenomena *b*, and glandular changes *c*. In a muscle *a* is large and *b* and *c* small; in a gland *a* may not occur as an active movement at all, although the cell may change slowly in volume; *b* is also small, but *c* is large; and in an electrical organ *a* is no doubt small (if it exist at all), *b* is very large, and *c* is small.

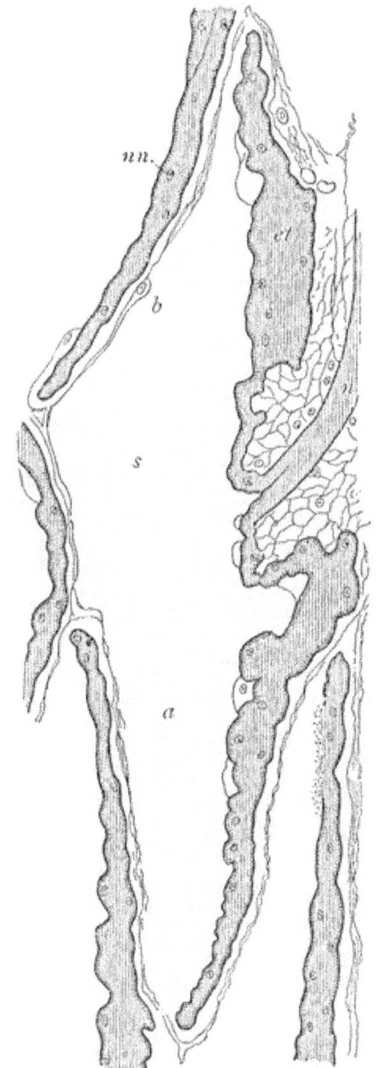

FIG. 81.—One of the lozenge-shaped spaces in the electric organ of malapterurus, magnified 200 diameters. *s*, space filled with fluid. To the left of *a* observe the electric tissue darkly tinted. Notice that it occurs on two sides of the lozenge-shaped space. See to the right of *b* the connective tissue wall of the space. *n n*, nuclei in electric tissue; *n*, nerve-fibre passing into electric tissue. The electric discs in this fish are epithelial and not muscular. The total number of discs, each of which is supplied with a nerve, is 2,000,000. All the nerves for the electric organs spring from two gigantic nerve-cells in the spinal cord, one for each lateral half. One nerve-process issues from each cell and, by dividing and subdividing, supplies each of the one million discs on one side of the body with a distinct nerve-fibre. The sum of the diameters of these nerve-fibres is very much greater than the diameter of the fibre that issues from the cell. It is evident, therefore, that the conducting matter of the nerve-fibres must increase in amount as we pass to the periphery of the body (p. 98).

Thus, in my opinion, all these phenomena are

but manifestations of the same essential process; they are all linked together, and as our knowledge of the nature of the molecular processes connected with life advances, we will be better able to explain and correlate such phenomena as contraction, secretion, and electrical action.

I have now the pleasure of showing you a large live gymnotus which has been kindly lent for demonstration by Mr. P. L. Sclater, the Secretary of the Royal Zoological Society. The fish, in charge of his keeper, is in this large tank. He is about four feet in length, and it is satisfactory to know that he has lived for seven years in the "Zoo," far away from his own Amazon, and that, with good feeding, he has nearly doubled in size. It is quite proper that the fish should first of all give a shock to a physiologist who is endeavouring to demonstrate his properties, so I seize hold of these handles while the keeper touches the fish with the ends of insulated wires. Ah! I have got a pretty smart shock, felt up to the elbow, like the discharge of a Leyden jar. Next we shall lead off a little to the galvanometer, making the instrument as insensitive as possible by pushing down the magnet; but you see how wildly the spot careers about on the scale when the keeper touches the fish. If we had time, we might cause the fish to stimulate a muscle in the frog-interrupter, and thus ring the bell. This is the method adopted by Professor du Bois Reymond in his investigations on electric fishes. If the keeper wishes it, we can easily fit up an arrangement in the Zoological Gardens by which the fish can ring him up at any time, say when he wants his dinner! I see by the keeper's face that this does not meet with his approbation: a bell at one end of a wire and a gymnotus at the other might almost be as troublesome an arrangement as a telephone in one's bedroom. We are much obliged to Mr. Sclater for lending the fish, and we hope the gymnotus will have a safe journey to his warm tank in the "Zoo."

Time warns me, however, that I must be drawing to a close. We have been trying in these lectures to get an insight into the hidden machinery connected with animal motion. Up to this point, we have only been discussing the mechanism of each individual wheel and pinion, and we have not considered the machine as a whole. It is impossible to discourse, with any degree of fulness, on this subject in the present course of lectures; but I shall content myself with alluding to one or two points of surpassing interest.

In the first place, these — the muscular mechanisms we have been considering—are controlled and regulated by the central nervous system. Each muscle is supplied by one or more nerves, and these originate in central nervous organs of great complexity, and regarding which much of our knowledge is singularly indefinite and unsatisfactory. We know, however, that there are two classes of movements—those that we make voluntarily and consciously, and those that we make involuntarily, and of which we may be either conscious or unconscious. We cannot make a voluntary movement without being conscious of so doing. An effort of will is always a conscious effort, and to speak of unconscious will, as some writers do, is, in my opinion, a very misleading mode of expression. What they mean, no doubt, is that certain movements may be made which are so purpose-like as to lead one to suppose that they are voluntary, and yet they may be made without the person or animal being conscious of making them.

Now, whether movements are voluntary or involuntary, they always require a nervous mechanism having the same structural type, although it may be more or less complicated by the necessities of the act to be performed. The simplest form of this mechanism is what we term in physiology a reflex mechanism. It consists of a centre, a sensory or afferent nerve, carrying impressions to the centre, and a motor or efferent nerve, carrying impressions from the centre to something in the circumference or periphery of the body. Thus, if we pinch the toe of a decapitated frog, it draws the leg away. The pinching irritates the sensory nerve, something, as you now know, travels along it to the nerve-centre, which, in this case, is in the frog's spinal marrow, and from the spinal marrow, after the lapse of a little time (the latent period in the marrow) a new impulse starts outwards along motor nerves to the muscles, reaches these, and causes them to contract. There are many varieties of these reflex acts. We may be quite unconscious of them, or we may feel the stimulus, and we may feel that we make a movement, and yet we may be unable to restrain the movement. Many reflex movements are beyond the control of the will when they have once been fairly set agoing. Thus, we cannot stop swallowing when the food has gone far enough back in the mouth and throat.

Now suppose we make the motions voluntarily: I wish to point out to you that the mechanism is still essentially of the same character. We usually speak as if we were free to make any movement we like, and when we know a little physiology we say the impulse begins somewhere in the brain and travels down nerves to the appropriate muscles. In a sense this is true, and yet it is not wholly true. We seem to act as if the mandate started in the brain, but this is because we miss the influences and impulses that called this mandate into activity. To start a mechanism that will produce well-ordered movement, as when I lift this book from the table, impulses or messages, whatever you like to call them, must first be transmitted from the body itself to the brain. Such messages may come to the brain by nerve-fibres from some organ of sense, it may be from the skin, or from the eyes, or from the muscles themselves, but they must be sent to the brain before the brain will send out messages along motor nerves to groups of muscles which are called into action, so as to perform a definite movement. If, from a disorder of the nerves, or of the nerve strands in the spinal cord that carry messages up to the brain, these messages cannot reach the brain, movements will either not be made at all, or if they are made, they are irregular, spasmodic, wanting in adjustment for a definite, purposive-like action. Thus sensory impressions come before and determine even so-called voluntary movements.

If this be the case, you will naturally inquire as to the mechanism by which certain messages sent to the brain are so arranged or transmitted as to call forth and transmit nerve-currents along certain specific nerve-fibres to certain specific muscles. We now get into a misty region in which we have only to grope our way. Analogies may help us a little. Is there something that answers the purpose of a telephonic exchange, in which a presiding genius, by putting in and taking out pegs, puts one part into connection with another? Or is there some kind of shunting-place, a sort of Clapham Junction, to which lines from all parts converge, and from which currents are sent here and there, according to the necessities that arise? Or is there at work something like the card of a Jacquard loom, by which all the threads are collected and arranged and transmitted, so that each takes its place in the complicated pattern of the woven web? All these analogies fail in giving a notion of the intricate phenomena that occur, and it must not be supposed that the nervous system works in the least like any one of the mechanisms I have alluded to. Still such arranging of the impulses does take place,—some think in the spinal cord itself, others in the cerebellum, others in the cerebrum, others in the nervous system as a whole,—and the result is exquisitely harmonised movement.

All these phenomena are undoubtedly connected with molecular movement. Such movements occur even in the brain itself, and there is little doubt they are also associated with all mental phenomena. It does not follow, however, that mental phenomena are the result of such movements alone. Wider knowledge strengthens the view that behind mental phenomena, and indeed behind all phenomena, there is something more than movements of matter and transformations of energy.

My task is now at an end. I have had great pleasure in delivering these lectures, because I felt that in endeavouring to interest you I was instructing myself. You have got a glimpse into the world of science, and I hope the glimpse will induce many of you to ask for more knowledge. Science is simply the truth about natural phenomena, so far as we can reach it. Some of you may become men of science, and you will probably advance much farther than we can do at present, and you will add to science, I hope, by your own work. The majority, however, will not follow scientific pursuits, but I trust this course of lectures will lead you always to keep a mind open for the reception of truth, from whatever quarter it may come, and that you will always cherish a lively sympathy with scientific men and with scientific progress.

In conclusion, let me thank Mr. Brodie of King's College and Mr. Heath of the Royal Institution for their valuable assistance; and let me also thank the Directors of the Royal Institution for giving the opportunity to a physiologist to represent his science in this lecture-theatre, as it may show how physiologists work and reason on the difficult problems with which they have to deal.

INDEX

THE END

Printed by R. & R. CLARK, *Edinburgh*

www.ingramcontent.com/pod-product-compliance
Lightning Source LLC
Chambersburg PA
CBHW021705210326
41599CB00013B/1525

* 9 7 8 3 3 3 7 2 9 1 1 5 0 *